The Internet *of* Things *in the* Cloud

A Middleware Perspective

Honbo Zhou

CRC Press
Taylor & Francis Group
Boca Raton London New York

CRC Press is an imprint of the
Taylor & Francis Group, an **informa** business

CRC Press
Taylor & Francis Group
6000 Broken Sound Parkway NW, Suite 300
Boca Raton, FL 33487-2742

International Standard Book Number: 978-1-4398-9299-2 (Hardback)

Library of Congress Cataloging-in-Publication Data

Zhou, Honbo.
The internet of things in the cloud : a middleware perspective / Honbo Zhou.
p. cm.
Includes bibliographical references and index.
ISBN 978-1-4398-9299-2 (hardcover : alk. paper)
1. Ubiquitous computing. 2. Cloud computing. 3. Middleware. 4. Embedded Internet devices. I. Title.

QA76.5915.Z68 2013
004.67'82--dc23 2012028834

Visit the Taylor & Francis Web site at
http://www.taylorandfrancis.com

and the CRC Press Web site at
http://www.crcpress.com

This book is dedicated to Yizheng, Robert, and Alexander, with love, for their understanding and support throughout the years.

Contents

List of Figures

List of Tables

Foreword

Imagine a world where everything around you, from smartphones, smart TVs, and light bulbs, to refrigerators, air conditioners, and motor vehicles, as well as countless instrumented objects including livestock, are connected, monitored, and sometimes, actuating and interacting among themselves, with or without human intervention, this is the world of the Internet of Things (IoT).

We have seen those applications today in real life such as GPS navigation systems for cars, mobile payment, smart city and intelligent building management, automatic meter reading, truck fleet locating, security surveillance and emergency management, natural resources and environment protection, air quality monitoring, space shuttle tracking, as well as scenarios in science-fiction movies. Many see this as the ultimate futuristic world, a world where ubiquitous smart devices and assets are connected to make human living easier and more convenient, and everything becomes smarter. With the introduction and development of IoT vision and technologies, this future world may be a lot closer to us than we think. Having worked on an Internet startup during the dotcom time right after receiving my PhD at Stanford, I feel that now is the time for IoT, just like 1999 was for the Internet and the web.

The possibilities of IoT/M2M are unlimited; however, the challenges are as enormous and pervasive both technologically, existing in every link of the value chain, and sociologically, affected by government regulations and user privacy concerns, and so on. For years, TCL Communication, which I joined right after the dotcom bust, has been known as one of the world's leading cellular phone makers (no. 3 in China and no. 7 worldwide) after the successful merge with Alcatel's cellular phone business. Now we are well positioned to take on the challenges with a synergic move together with the entire TCL Group, which has broader IoT/M2M–related product lines, and capture an even bigger slice of the trillion-dollar pie.

With the advent of cloud computing and IoT, along with the convergence of all types of intelligent devices, I believe mobile Internet is ready to take off. M2M terminals including smartphones are going to play a pivotal role in the mobile Internet revolution. Smart M2M devices are becoming the gateways to the converging fixed and mobile Internet just like personal computers for the Internet years ago. Currently, almost all the major Internet players, such as Google, Amazon, Microsoft, and even Facebook, are entering the smartphone business. Apple's success with iTunes, AppStore, and iCloud make it a competitive Internet (fixed and mobile) player. Almost all of the Telco operators such as AT&T, Orange, Verizon, and NTT DoCoMo, to name a few, have developed M2M business strategies and made substantial progress since 2004 in the M2M market. Telco equipment players such as Cisco also saw and embraced M2M or IoT as the next big thing. Mobile phone chip maker Qualcomm has quietly become the largest fleet telematics services provider (TSP) with 600,000 vehicles receiving its services. According to e-Principles, a market research organization, the number of cellular M2M devices surpassed the number of cell phones in West Europe in 2010.

At TCL Communication, we also realized the importance of a unified software platform with cloud services built on top for smartphones, smart TVs, smart home appliances, healthcare monitors, and other M2M devices. The TCL T-Cloud strategic blueprint was announced as a unified foundation for a variety of M2M/IoT vertical applications. A number of projects are in development to embrace the market potentials and take advantage of the great opportunities. This entire value chain and the associated technologies, as well as the enormous and ubiquitous application landscape, are a vast topic that encompasses many different subjects. This book brings timely, mind-provoking, and comprehensive materials to help you have a better understanding of the IoT/M2M technological and business landscape on top of cloud computing, and shape your business strategies. I also believe this book, which I highly recommend, is the first on the market that covers almost all of the related subjects.

George (Aiping) Guo, PhD
CEO of TCL Communication Technology Holdings Ltd.
Senior Vice President of TCL Group

Preface

IoT was embraced by countries worldwide, especially in Asia and Europe; for example, IoT is part of national strategy in China; Japan has been promoting U-Japan since 2004; the EU is aiming to "lead the way" in the transformation to Web 3.0 with the Internet of Things; and so on. Although almost all of the new concepts in the IoT domain including the Internet of Things itself as well as other related terms such as M2M (Machine to Machine), CPS (Cyber-Physical Systems), Smarter Planet, Smart Grid, started their life in the United States and became buzzwords worldwide; the IoT term is not yet a catchphrase in the U.S. People in the U.S. seem busy enough with the social networking hype (Facebook, Twitter, etc.) and regard the Internet of Things as the Internet of Somebody-else's Things. Many people also believe that there is nothing new technologically about IoT. It's true that all of the technologies and most of the applications that enable and constitute IoT existed long before anyone ever began to talk about IoT. Just like the Internet and web, the Internet of Things is a revolutionary way of architecting and implementing systems and services based on evolutionary changes. The realization of the IoT vision brings ICT technologies closer to many aspects of real-world life instead of virtual life, and therefore has greater implications and sociological impacts in a world facing serious issues such as global warming, environment protection, and energy saving. It's hoped that this book would help bring awareness for more people to know IoT and join forces to boost faster development of IoT.

This book provides a panoramic view of the IoT landscape and focuses on the overall technological architecture and design of a tentatively unified IoT system underpinned by different cloud computing paradigms from a middleware perspective. It is based on the author's two previous best-selling books (in Chinese) on IoT and cloud computing and more than 20 years of hands-on software/middleware programming and architecting experience in the United States. The author worked at Oak Ridge National Laboratory, IBM, BEA Systems, and Silicon Valley startups such as Doubletwist. While at Doubletwist, the author led a team that created a COW (cluster of workstations) or grid computing (now cloud computing) system (and a Software as a Service [SaaS] portal on top of it) that accomplished the complete annotation of the entire human genome for the first time in the world. This accomplishment was reported by the *San Francisco Chronicle* (http://www.sfgate.com/business/article/The-Gene-Team-High-tech-gurus-biologists-unite-3304818.php) and CNN, and media in Asia and Europe.

Most directly, this book is based on the author's research and development endeavors on the ^{ez}M2M middleware of the TongFang Co. Ltd. (the second largest system integrator and IT services provider in China) platform and more than 30 IoT vertical application suites on top of it since 2003. This platform has received more than 20 awards and recognitions in China and has been used in more than 800 projects including the radio-frequency identification (RFID)-based ticket management system for the 2008 Olympic Games, the M2M platform for China Mobile and e-Logistics vehicle tracking system, the national emergency management system, the building management system for the central television tower of China, and the smart building energy efficiency management system for the city of Beijing. The ^{ez}IBS building management application suite product has maintained the number one market share position since 2006, and is included in textbooks for college students of related majors in China. The author expresses his gratitude to all the members of the ^{ez}M2M R&D team.

This book is comprised of three sections. The first section describes the concept of Internet of Things. Other related concepts along with its development, and a number of important vertical IoT applications, as examples, are also demonstrated. The four pillars of IoT are introduced based on the author's extensive and exhaustive research and industry practices, and it is believed that those four paradigms represent the most comprehensive and holistic clarification and categorization that cover all of IoT's nuts and bolts. The three-layer value chain of IoT is described in the last chapter of Section I.

The Web of Things (WoT) is a better term to describe what the Internet of Things is meant to be, just like the World Wide Web is based on the Internet. It is more about the so-called grand integration and applications rather than the ubiquitous networks and the devices and sensors. So the middleware, just like the three-tiered application servers for the web, and the associated data formats, such as HTML- and XML-based data representations for EAI (Enterprise Application Integration) and B2B, play a pivotal role in the entire IoT value chain. The second section of the book focuses on middleware, the glue and building blocks of the holistic IoT system in every layer of the architecture. A comprehensive overview of all sorts of middleware and their roles in the four IoT pillar systems is presented in the first chapter of Section II. The data formats and protocols for all the four pillar IoT applications are summarized, and the possibility of creating a unified IoT data format and protocol standard for the four pillar segments is investigated in Chapter 6. The last chapter of Section II investigates the possibility of creating a unified IoT middleware architecture based on currently existing research efforts on standardization, such as IoT-A (Internet of Things Architecture) and ETSI (European Telecommunications Standards Institute) M2M functional architectures and a number of commercial products.

The third section of the book discusses cloud computing and IoT as well as their synergy based on the common background of distributed processing. The fundamentals of cloud computing are discussed in Chapter 8. The MAI (Machine to Machine Application Integration) similar to EAI integration inside a firewall, and the XaaS (Everything as a Service), similar to B2B/B2C based on SOA (Service Oriented Architecture) over the Internet, paradigms for IoT/WoT integration are introduced and discussed, and a comprehensive unified IoT framework specification is proposed and explained in the final chapter of the book.

The Internet of Things is a vast and dynamic territory and is evolving at a rapid pace. Books that offer a comprehensive and holistic view are not yet seen on the market. This book attempts to be a comprehensive guide to IoT technologies and system architectures. However, it is more like a research report that introduces a few new propositions based on the author's R&D endeavors rather than a textbook. The audience for this book could be software engineers, architects, post-graduate students, researchers, or anyone who wants to know more about IoT, especially its software technologies and system architectures.

The author can be found on Facebook, Twitter, and LinkedIn. He can also be reach at honbozhou@gmail.com. Comments and suggestions as well as criticisms are welcome.

Author

Honbo Zhou, PhD, is currently the general manager of Foton Fleet Telematics Co. Ltd. He was chief software scientist of TongFang Co. Ltd., executive director of the board and chief technology officer of Technovator Pte. Ltd., and chief operating officer of TongFang Software Co. Ltd. Dr. Zhou worked as a research associate on grid computing at Oak Ridge National Laboratory after receiving his PhD in computer science from the University of Zurich in 1993. He also worked at IBM, BEA Systems (now Oracle), and other companies in the United States as a senior engineer or manager. He participated in the ASCI Blue Pacific project, building the world's fastest supercomputer in 1996 while at IBM as a software team lead and coordinator for its job scheduler. He masterminded and built a high-performance/cloud-computing system that accomplished the complete annotation of the human genome for the first time in the world while working at a startup in Silicon Valley, which was reported by media such as the *San Francisco Chronicle* and CNN. He has been one of the pioneers of the Internet of Things and machine-to-machine computing, leading a team of 100-plus developers to build TongFang's flagship ezM2M Middleware Platform for dozens of vertical IoT applications since 2003, and he has authored two related books in Chinese. He is a frequently invited speaker and evangelist of IoT and cloud computing, adjunct professor of several universities, and vice president of the Middleware Association, and he is a member of other related professional associations in China.

THE INTERNET OF THINGS

Chapter 1

The Third ICT Wave

1.1 Rise of the Machines

Over the past decades, billions of people have hooked themselves up to the Internet via the computer, and more recently mobile devices such as smartphones. This communication revolution is now extending to objects as well as people. Machine-to-machine (M2M) communication has long been predicted, and now it is rushing into the present. According to Parks Associates, the number of smartphones (excluding feature phones) worldwide is expected to top 1.1 billion in 2013. However, this is just the tip of the iceberg. Smart grid devices will reach 244 million; e-readers and tablets will be 487 million; networked office devices, 2.37 billion; networked medical devices, 86 million; connected automobiles, 45 million; connected appliances, 547 million; connected military devices, 105 million; information technology (IT) system devices, 431 million; connected supervisory control and data acquisition (SCADA)/industry automation devices, 45 million; and other connected consumer electronic devices minus smartphones, e-readers, and tablets will reach a whopping 5+ billion and counting.

"Rise of the machines" became a popular catchphrase after *Terminator 3: Rise of the Machines*, a 2003 science-fiction action film directed by Jonathan Mostow and starring Arnold Schwarzenegger. The movie demonstrates the power of machines or robots that could potentially overpower human beings.

During the first decade of the twenty-first century, big U.S. defense budgets financed the deployment of thousands of service robots, including unmanned aerial and underwater vehicles, to Iraq and Afghanistan. *IEEE Spectrum* [1] estimated a million industrial robots toiling around the world in 2008, and Japan is where they're the thickest on the ground. In 2011, the world's industrial robot population was estimated to be 1.2 million. Also, according to the Frankfurt-based International Federation of Robotics, the service robot market is expected to double in size by 2013 from 2011 [2].

A robot is a kind of tightly coupled cyber-physical system (CPS) [4,165]. A CPS (Figure 1.1) is an embedded sensor network and control system featuring a tight combination of, and coordination between, the system's computational and physical elements. Cyber-physical systems or robots can be found in areas as diverse as aerospace, automotive industry, chemical

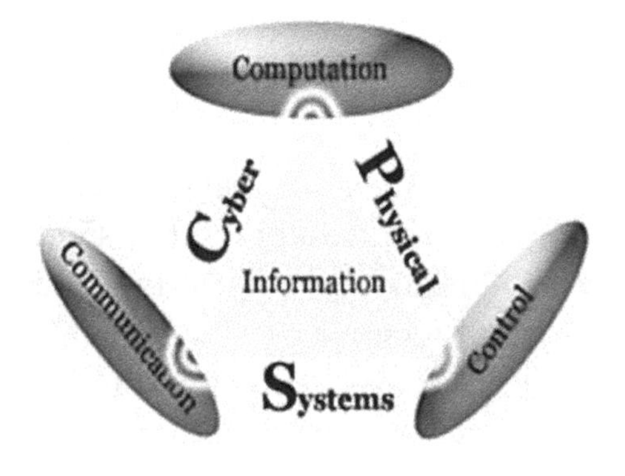

Figure 1.1 Cyber-physical system (CPS).

processes, civil infrastructure, energy, healthcare, manufacturing, transportation systems, entertainment, and consumer appliances. A real-world example of such a system is the Massachusetts Institute of Technology (MIT) CarTel project where a fleet of taxis collects real-time traffic information in the Boston area. Together with historical data, this information is then used for calculating the fastest route for a given time of the day.

The U.S. National Science Foundation (NSF) has identified cyber-physical systems as a key area of research, proposed by Helen Gill at the High Confidence Software and Systems conference [28] in 2008. In 2007, the President's Council of Advisors on Science and Technology listed CPS as one of the top eight key technologies of the future, and a $4 billion budget was allocated for the Networking and Information Technology Research and Development [29] project. The expectation is that in the coming years, ongoing advances in science and engineering will improve the link between computational and physical elements, dramatically increasing the adaptability, autonomy, efficiency, functionality, reliability, safety, and usability of cyber-physical systems.

The power of machines has experienced rapid development, first through the steam-engine technology based industrial revolution and then the second electrical, oil-powered internal combustion engine industrial revolution. Along with the rise of the power of machines comes the exponential rise of the number of machines during the ongoing third industrial revolution of the Internet-based information age. The past three decades have seen extraordinary growth in the number and choice of electrical and electronic machines or devices (Figure 1.2) [3].

The so-called Internet of Things (IoT), together with cloud computing, is, after the modern computer (1946) and the Internet (1972), the world's third wave of the information and communications technology (ICT) industry. Gordon Bell's law

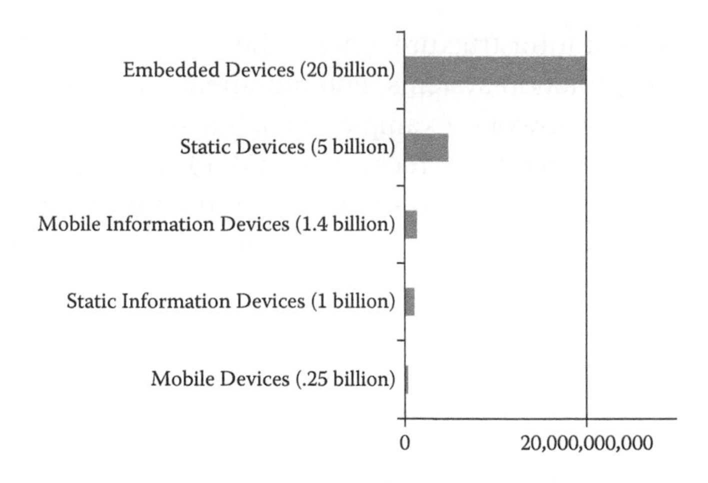

Figure 1.2 Number of intelligent devices.

says that "roughly every decade a new, lower priced computer class forms based on a new programming platform, network, and interface resulting in new usage and the establishment of a new industry" [271]. Bell predicted that home and personal area networks will form starting from 2010.

Also, in 2002, Sun's chief technology officer Greg Papadopoulos indicated that the first Internet wave consisted of an "Internet of computers" and the second wave, which we are currently in, is an "Internet of Things that embed computers." The third Internet wave, which is an "Internet of Things," consists of physical objects like thermostats, switches, packages, and clothes.

So far, our view of the Internet has been human-centric. It is quite likely that sooner or later the majority of items connected to the Internet will not be humans, but things. The IoT will primarily expand communication from the 7 billion people around the world to the estimated 50 to 70 billion machines. This means significant opportunities for the telecom industry to develop new IoT "subscribers" that substantially surpass the number of current subscribers based on population.

This advancement signifies a massive shift in human development, from an electronic society to a ubiquitous society in which everything is connected (for example, the sensor in Nike+ shoes sends information to an iPod application [192]) and everything can be accessed anywhere. Supported by IPv6 and eventually the Future Internet Architecture, the IoT would have the potential of connecting the 100 trillion things that are deemed to exist on Earth [17].

Recent developments predict that we will have 16 billion connected devices by the year 2020 [5], which will average out to six devices per person on earth. Devices like smartphones and M2M or thing-to-thing communication will be the main drivers for further development.

Cisco's Dave Evans has posted a great infographic (http://blogs.cisco.com/news/the-internet-of-things-infographic/) showing that communicating *things*, essentially embedded sensors, have already outstripped the number of communicating *homo sapiens* in 2010. Future historians will probably look back at 2010 as the year when Internet-connected devices like digital picture frames, web-connected global positioning system devices, and broadband TVs came online in greater numbers than new human subscribers. Electricity meters, dishwashers, refrigerators, home heating units, and several other objects with tiny sensors are next in line.

By 2015, wirelessly networked sensors in everything we own will form a new web. But it will only be of value if the terabyte torrent of data it generates can be collected, analyzed, and interpreted [6]. The first direct consequence is the generation of huge quantities of data from physical or virtual objects that are connected. As a result, consumer-device-related messaging volume could easily reach between 1,000 and 10,000 per person per day [7,8].

As a key aspect of the next-generation Internet, the Internet of Things is expected to have a dramatic impact on almost all sectors of the web-based service economy. It will enable tremendous efficiency gains, especially in the transportation, retail, manufacturing, logistics, and energy sectors. The world market for Internet of Things–related technologies, products, and applications alone will increase significantly from $2 billion today to more than $11.5 billion in 2012, with average annual growth rates of almost 50 percent [269]. More aggressive forecasts predict a market volume of more than $27 billion in 2011 [270]. Forrester Research also predicts that the number of objects connected to the IoT will be 30 times the number of people connected to the Internet by 2020. IoT is a trillion-dollar industry.

1.2 The IoT Kaleidoscope

Although the concept of IoT was expressed in the form of "computers everywhere" by professor Ken Sakamura (University of Tokyo) in 1984 and "ubiquitous computing" by Mark Weiser (Xerox PARC) in 1988, the phrase *Internet of Things* was coined by Kevin Ashton (Procter & Gamble) in 1998 [9] and developed by the Auto-ID Center of MIT from 2003. Ashton then described the IoT as "a standardized way for computers to understand the real world." MIT has also contributed significant research in this field, notably Things That Think consortium at the Media Lab and the CSAIL effort known as Project Oxygen. Other major contributors include Georgia Tech's College of Computing, New York University's Interactive Telecommunications Program, University of California at Irvine's Department of Informatics, Microsoft Research, Intel Research and Equator, and Ajou University UCRi and CUS.

The concept of IoT has since become popular through the radio-frequency identification (RFID) Auto-ID Center's six research labs in the United States, United Kingdom, Australia, Switzerland, Japan, and China. It refers to uniquely identifiable objects and their virtual representations in an Internet-like architecture. Although the idea is simple, its application is powerful. If all objects of daily life were equipped with radio tags, they could be identified and inventoried by computers

[10,11], and daily life on our planet could undergo a drastic transformation [12].

In the International Telecommunication Union (ITU) Internet report of 2005 [13] and the EPOSS's (European Technology Platform on Smart Systems Integration) IoT 2020 report [22], however, the concept of IoT was further extended to cover a plethora of technologies, applications, and services beyond RFID and the aforementioned CPS, which will enhance quality of life while providing new revenue opportunities for a host of enterprises. The Internet as we know it is transforming radically, from an academic network in the 1980s and early 1990s to a mass-market, consumer-oriented network. Now, it is set to become fully pervasive, connected, interactive, and intelligent. Real-time communication is possible not only by humans but also by things at any time and from anywhere.

Over two decades ago, the late Mark Weiser of Xerox PARC developed a seminal vision of future technological ubiquity—one in which the increasing *availability* of processing power would be accompanied by its decreasing *visibility*. As he observed, "the most profound technologies are those that disappear ... they weave themselves into the fabric of everyday life until they are indistinguishable from it" [272]. Weiser is widely considered to be the father of *ubiquitous computing*, a term he coined in 1988.

According to Weiser, "Ubiquitous computing names the third wave in computing, just now beginning. First were mainframes, each shared by lots of people. Now we are in the personal computing era, person and machine staring uneasily at each other across the desktop. Next comes ubiquitous computing, or the age of calm technology, when technology recedes into the background of our lives." *Pervasive computing* is a similar term used by IBM's former chief executive officer (CEO) Louis Gerstner in 1996, when I joined IBM as a software programmer doing job-scheduling software development in the SP PowerParallel Division that built the world's fastest supercomputer at the time, ASCI-Blue Pacific.

Just like CPS, ubiquitous computing is synonymous with or closely related to IoT. About a dozen other terms are synonymous with or closely related to IoT, which can be regarded as an umbrella word to cover the technologies and applications that these terms or phrases describe. A comprehensive (but not complete due to the ever-changing nature of technology developments) collection of those terms and phrases is listed and explained in the following paragraphs.

M2M (machine-to-machine) refers to technologies that allow both wireless and wired devices to communicate with each other or, in most cases, a centralized server. An M2M system uses devices (such as sensors or meters) to capture events (such as temperature or inventory level), which are relayed through a network (wireless, wired, or hybrid) to an application (software program) that translates the captured events into meaningful information (such as the statistics of a vehicle's usage in OnStar). M2M communication is a relatively new business concept, born from the original telemetry technology, utilizing similar technologies but modern versions of them.

Telemetry is a technology that allows remote measurement and reporting of information. Systems that need external instructions and data to operate require the counterpart of telemetry, telecommand. Many modern telemetry systems take advantage of the low cost and ubiquity of GSM networks by using SMS to receive and transmit telemetry data. Telemetry has unlimited applications in many fields including meteorology, space science, agriculture, water management, defense, resource exploration, rocketry, medicine, and so on.

A wireless sensor network (WSN) consists of spatially distributed autonomous sensors to monitor physical or environmental conditions, such as temperature, sound, vibration, pressure, motion, or pollutants, and to cooperatively pass their data through the network to a main location. The more modern networks are bidirectional, becoming wireless sensor and actuator networks (WSANs) enabling the control of sensor activities.

In 2008, IBM's CEO Sam Palmisano outlined a new agenda for building a "smarter planet" during a speech [14] at the Council on Foreign Relations. The IBM initiative seeks to highlight how forward-thinking leaders in business, government, and civil society around the world are capturing the potential of smarter systems to achieve economic growth, efficiency, sustainable development, and societal progress. Examples of smarter systems include smart grids, water management systems, solutions to traffic congestion problems, and greener buildings. These systems have historically been difficult to manage because of their size and complexity. But with new ways of monitoring, connecting, and analyzing the systems, business, civic, and nongovernmental leaders are developing new ways to manage these systems. The IBM initiative was embraced by President Obama [15] and Smarter Earth became a U.S. government initiative. A $3.4 billion grant for smart grid was announced by President Obama later in 2009 [16]. Smart Grid is poised to "change" the energy efficiency management landscape.

In November 2008, *Time* magazine listed the IPSO (Internet Protocol for Smart Objects) Alliance and the Internet of Things among the most important innovations of 2008. Also in 2008, the U.S. National Intelligence Council published a report titled, "Disruptive Civil Technologies: Six Technologies with Potential Impacts on U.S. Interests out to 2025." These technologies are biogerontechnology, energy storage materials, biofuels and bio-based chemicals, clean coal technologies, service robotics, and the Internet of Things. With regard to the Internet of Things, it stressed the following:

> By 2025 Internet nodes may reside in everyday things—food packages, furniture, paper documents, and more. Today's developments point to future opportunities and risks that will arise when people can remotely control, locate, and monitor even the most mundane devices and articles. Popular demand combined with technology advances could drive

> widespread diffusion of an Internet of Things (IoT) that could, like the present Internet, contribute invaluably to economic development and military capability. [194]

Many U.S. companies are involved and playing important roles, with information technology (IT) giants such as IBM focusing on applications, Cisco on infrastructures, and so on.

Some experts predict that the IoT will help tackle two of the biggest problems facing mankind today: energy and healthcare. Currently buildings waste more energy than they use effectively, but we will be able to cut this waste down to almost nothing. Currently we make visits to our general practitioner twice a year, at most, but we will be able, thanks to a few sensors discreetly attached to our body, to continuously monitor our vital functions. Those two issues are among the top of President Obama's agenda. In 2009, Obama reiterated [20] his commitment to healthcare reform and stood firm on his assertion that healthcare IT must to be at the crux of reform. *Telehealth* or *telemedicine* are terms that are related to IoT.

In *Shaping Things*, the latest book by world-renowned science-fiction writer and futurist Bruce Sterling [27], ideas are outlined for *spime*, a word the author coined in 2004. A spime is, by definition, the protagonist of a documented process. It is a historical entity with an accessible, precise trajectory through space and time. It can also be a form of ubiquitous computing that gives smarts and searchability to even the most mundane of physical products. Imagine losing your car keys and being able to search for them with Google Earth. The three facets of spime that are relevant to IoT are as follows:

- Small, inexpensive means of remotely and uniquely identifying objects over short ranges
- A mechanism to precisely locate something on Earth
- A way to mine large amounts of data for things that match some given criteria

More recent ideas have driven the IoT toward an all-encompassing vision to integrate the real world into the Internet—the real-world Internet (RWI) [163]. RWI and IoT are expected to collaborate with other emerging concepts such as the Internet of services (IoS), and the building block of parallel efforts such as the Internet of energy (IoE) is expected to revolutionize the energy infrastructure by bringing together IoS and IoT/RWI. It is clear that the RWI will heavily impact the way we interact in the virtual and physical worlds, overall contributing to the effort of the future Internet.

Other terms or phrases that are relevant but more academic include sentient computing, haptic computing, physical computing, ambient intelligence, context-aware computing [18], things that think, autonomic computing, machine that talks, everyware [19], network embedded devices [170], domotics, and so on.

As you can see, the IoT-related terms come in different shapes and forms; a kaleidoscope-like picture [74] of IoT-relevant terms and phrases is shown in Figure 1.3. Despite the technology existing in its various forms, IoT comprises of a number of separate technologies that need to be mixed and matched in the appropriate manner to enable a broad market deployment.

Figure 1.3 IoT-related terms.

1.3 Defining Internet of Things

The IoT is a concept that has received considerable and significant attention and support within the European Commission (EC) with respect to strategic developments for ICT and the Information Society. Viviane Reding, vice president of the EC, in a speech to the Future of the Internet initiative of the Lisbon Council identified the IoT as an important driver for the Internet of the future [5].

An EC communication to the European Parliament, the Lisbon Council, the European Economic and Social Committee, and the Committee of the Regions entitled "Internet of Things: An Action Plan for Europe" was adopted on June 18, 2009, and reinforces the commitment to the concept and its importance for Europe, quoting the following in its conclusions [11]:

> Internet of Things (IoT) is not yet a tangible reality, but rather a prospective vision of a number of technologies that, combined together, could in the coming 5 to 15 years drastically modify the way our societies function. By adopting a proactive approach, Europe could play a leading role in shaping how IoT works and reap the associated benefits in terms of economic growth and individual well-being, thus making the Internet of Things an Internet of Things for people.

In China, a number of significant public speeches about IoT were delivered in the second half of 2009. On August 7, Chinese Premier Wen Jiabao made a speech in the city of Wuxi calling for the rapid development of Internet of Things ("Sensing China" was the term in Chinese he used to refer to what the IoT technologies should be used for) technologies (Figure 1.4). Rapidly, an "IoT wave" spread across the nation. IoT became a buzzword instantaneously in China. Government officials at all levels, as well as the rank and

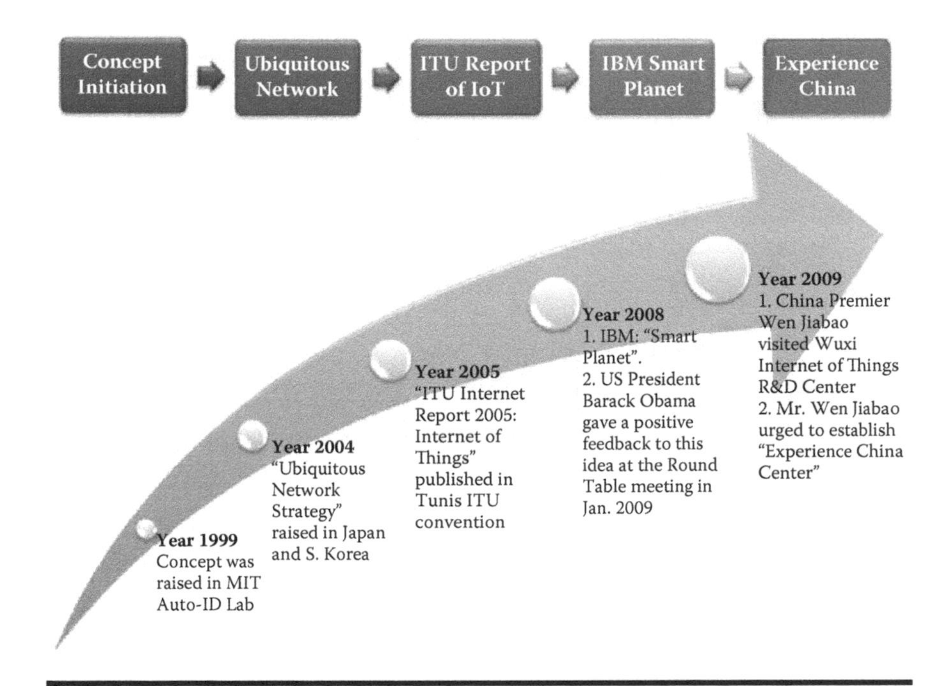

Figure 1.4 IoT development.

file, began trying to understand what the Internet of Things is. More than 60 books on this topic have been published in China since 2010.

Wen Jiabao followed up with another speech on November 3 at the Great Hall of the People in Beijing, in which he called for breakthroughs in wireless sensor networks and the Internet of Things. IoT was written into Premier Wen's "government work report" during the National People's Congress and the Chinese People's Political Consultative Conference in 2010, and the development of IoT industry became a national strategy. As a consequence, IoT was also written into the nation's "Twelfth Five-Year" plan in 2011. In response to the central government's initiative, over 60 related alliances and consortia were formed throughout China. Since 2009 [167], IoT has almost become a household buzzword (Figure 1.5).

The Chinese believe that China missed the first two waves of the ICT (information and communications technology) industry developments. Now, though, China may be well

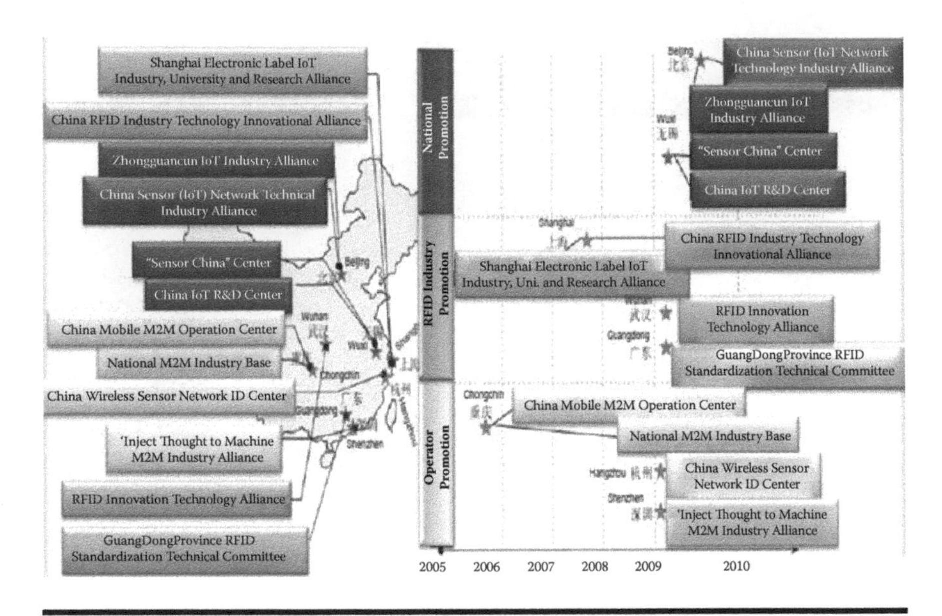

Figure 1.5 IoT development in China.

poised to take part or even lead in the third IoT wave based on the leapfrogging theory [23]. The fact is that China has the largest customer base, newer ICT infrastructure, and a determined and centralized government that has the almighty power of allocating and consolidating (top-down instructional planning versus bottom-up endless democratic debating/hearing in the Western world) national resources. IoT is more about infrastructure at the current stage than about income-generating, innovative business models. Figure 1.6 shows a SWOT (strengths, weaknesses, opportunities, and threats) analysis of China's IoT initiative and development [74].

In Japan and Korea, the buzzword is *ubiquitous* computing or the letter u as a prefix to a number of words such as u-Korea, u-Japan, u-city, u-home, u-tourism, u-business, u-defense, u-government, to name a few, rather than IoT, but these refer to the same thing. The u-words are sprinkled all over presentations, descriptions, and reports. There is a ubiquitous economy and the ubiquitous society; to sum everything up, there is u-life. The u-fever started around 2004 when the

SWOT Analysis of China's Advantage of IoT Development

	Positive	Negative
Internal Factors	**Strength** • IoT is a national strategy • Power of centralized government • Fast economic growth • Nationwide support • Largest customer base • World-class newer infrastructure • Lower labor costs • Lots of money to invest	**Weakness** • Underdeveloped middleware industry • Experience in standardization • Underdeveloped sensor technologies • Software recognition and capabilities • Top-notch talents • International business development • Repeat low-tech works • Waste of resources
External Factors	**Opportunities** • At the same start-line with peers • Leapfrogging advantages • Standardization opportunities • Economic stagnation outside • Lack of focus in US • Hard to consolidate in Western nations • Higher labor costs in Western nations	**Threats** • Better international products • Not owner of major standards • Lack of industry giants • Loss of talents to foreign competitors • National security concerns

Figure 1.6 SWOT analysis of China's IoT.

term M2M become popular in the United States. Before the *u* (ubiquitous) era was the *e* (electronic) era. The *e* era is concerned with the acceptance of digital communication for legally binding information. The *u* era proceeds to include objects, not humans only, in the circle of information producers and consumers [21].

Although the terms *Internet of Things* or *ubiquitous computing* were coined by Americans, they didn't become as popular in the United States. As mentioned before, the slogans "smarter planet" or "wisdom of Earth" were proposed by IBM, which again seized the Zeitgeist and told the right story at the right time to the right people in the depth of economic recession and financial crisis as well as climate change and global warming challenges. These were adopted by President Obama, who is trying find a dotcom-like innovation that could catalyze new markets for sustainable growth and save the economy. "Smarter planet" and "wisdom of Earth" refer to almost the same thing as IoT or u-life. Terms like *smart earth*, *smart grid*, *smart home*, *smart city*, and so on are more widely used in the United States, which indicates that the U.S. people as a whole think in terms of "smarter."

This is perhaps a coincidence but it isn't a joke. People in the United States seem more practical and tend not to follow what the government or an authority (such as EC) says. Some in the United States think IoT is the Internet of European things (and jokingly call the European Parliament the "Parliament of Things" [164]): a fiasco, or a big concept with no substance. That's probably why IoT is not a buzzword in the United States, like cloud computing, software as a service (SaaS), SOA, and others. Instead, *connectivity* is becoming a more popular term after M2M that refers to the same thing as IoT but more to the "real matter" and innovative new business model creations. However, we should not forget that the Europeans invented the Web. It seems that they are now on track to make the Internet of European Things into the Internet of Real Things, according to Viviane Reding [162].

Nevertheless, the *Internet of Things* is arguably still the most comprehensive term to describe the all-inclusive contents that the aforementioned terms and phrases refer to. This book is trying to raise awareness and acceptance of the term *Internet of Things* in the United States as well as elsewhere in the world. But what is the Internet of Things?

Due to the multifaceted, all-inclusive nature and scope of the Internet of Things, it's almost impossible to have a definition that everyone agrees on. IoT means different things to different people, just like the story about the six blind men and the elephant.

Below are a few definitions of the Internet of Things, and most come from Europe.

- CASAGRAS's (Coordination and Support Action for Global RFID-related Activities and Standardization) IoT definition:

  IoT is a global network infrastructure, linking physical and virtual objects through the exploitation of data capture and communication capabilities. This infrastructure includes existing and evolving Internet and network developments. It will offer specific object-identification, sensor and connection capability as the basis for the development of independent cooperative services and applications. These will be characterized by a high degree of autonomous data capture, event transfer, network connectivity and interoperability [24].

- SAP's IoT definition:

  IoT is going to create a world where physical objects are seamlessly integrated into the information network, and where the physical objects can become active participants in business processes. Services are available to interact with these "smart objects" over the Internet, query and change their state and any information associated with them, taking into account security and privacy issues [25].

- EPoSS's (the European Technology Platform on Smart Systems Integration) IoT definition:

  The network formed by things/objects having identities, virtual personalities operating in smart spaces using intelligent interfaces to connect and communicate with the users, social and environmental contexts [22].

- CERP's (Cluster of European RFID Projects) IoT definition:

  Internet of Things is an integrated part of Future Internet and could be defined as a dynamic global network infrastructure with self configuring capabilities based on standard and interoperable communication protocols where physical and virtual "things" have identities, physical attributes, and virtual personalities and use intelligent interfaces, and are seamlessly integrated into the information

> network. In the IoT, "things" are expected to become active participants in business, information and social processes where they are enabled to interact and communicate among themselves and with the environment by exchanging data and information "sensed" about the environment, while reacting autonomously to the "real/physical world" events and influencing it by running processes that trigger actions and create services with or without direct human intervention. Interfaces in the form of services facilitate interactions with these "smart things" over the Internet, query and change their state and any information associated with them, taking into account security and privacy issues [26].

The definition of IoT depends very much from the aspect or angle examined. The aforementioned definitions are mostly from an RFID point of view. A comprehensive, all-inclusive view should be sought.

- IoT definition or statement of this book (Figure 1.7):

  The Internet of Things is a plethora of technologies and their applications that provide means to access and control all kinds of ubiquitous and uniquely identifiable devices, facilities, and assets. These include equipment that has inherent intelligence, such as transducers, sensors, actuators, motes [179], mobile devices, industrial controllers, HVAC (heating, ventilation, and air-conditioning) controllers, home gadgets, surveillance cameras, and others, as well as externally enabled things or objects, such as all kinds of assets tagged with RFID, humans, animals, or vehicles that carry smart gadgets, and so forth. Communications are via all sorts of long- and

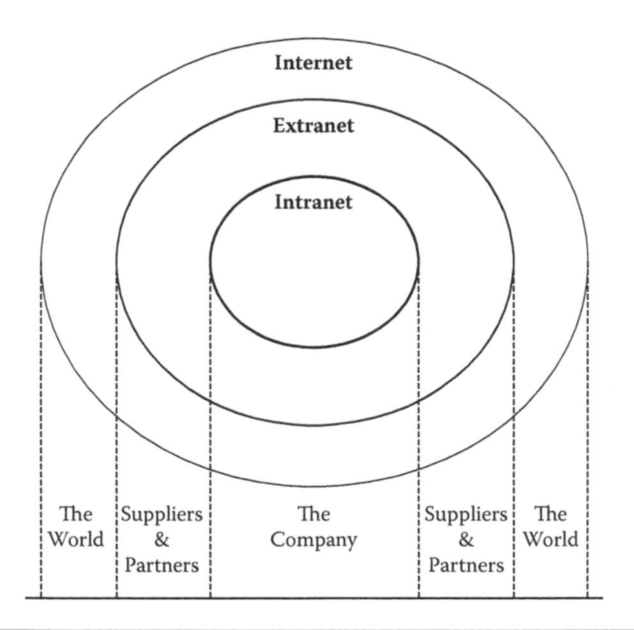

Figure 1.7 Intranet/Extranet/Internet.

short-range wired or wireless devices in different kinds of networking environments such as Intranet, extranet, and Internet that are supported by technologies such as cloud computing, SaaS, and SOA and have adequate privacy and security measures, based on regulated data formats and transmission standards. The immediate goal is to achieve pervasive M2M connectivity and grand integration and to provide secure, fast (real time), and personalized functionalities and services such as (remote) monitoring, sensing, tracking, locating, alerting, scheduling, controlling, protecting, logging, auditing, planning, maintenance, upgrading, data mining, trending, reporting, decision support, dashboard, back office applications, and others. The ultimate goal is to build a universally connected world that is highly productive, energy efficient, secure, and environment friendly.

1.4 IoT: A Web 3.0 View

The Internet (network) and the web (application) are two sides of a coin. The Internet was invented by Vinton Cerf in 1973, and the invention of the web in 1989 was credited to Tim Berners-Lee and later caught worldwide attention by Marc Andreessen's Mosaic web browser in 1992. The Internet (hardware) is the infrastructure and the web (software) is the application everybody uses. Just like the Internet revolution, in the Internet of Things, web-based applications and software (the supporting data representation and middleware) are the keys.

McKinsey [36] summarized the key application functionalities of IoT systems:

1. Information and analysis
   a. Tracking behavior
   b. Enhanced situational awareness
   c. Sensor-driven decision analytics
2. Automation and control
   a. Process optimization
   b. Optimized resource consumption
   c. Complex autonomous systems

According to Harbor Research, the web-based applications, systems, and networked services of smart systems or IoT are expanding more rapidly than the hardware and infrastructure [37]. This means the software (middleware and web-based integrated applications) market will play a pivotal role in the IoT business.

As is well known, Web 1.0 is about publishing and pushing content to the users. It's mostly a unidirectional flow of information. The shift from Web 1.0 to Web 2.0 can be seen as a result of technological refinements as well as the behavior change of those who use the World Wide Web, from publishing to participation, from web content as the outcome of large up-front investment to an ongoing and interactive process. Web 2.0 is about two-way flow of information and is associated with web applications that facilitate participatory information sharing, interoperability, user-centered design, and collaboration. Example applications of Web 2.0 include blogs, social networking services (SNSs), wikis, mashups, folksonomies, video-sharing sites, massive multiplayer online role-playing games, virtual reality, and so on.

Enterprise 2.0 is the use of Web 2.0 technologies within an organization to enable or streamline business processes while enhancing collaboration (Figure 1.8). It is the extension of Web 2.0 into enterprise applications. IoT technologies and applications can be integrated into Enterprise 2.0 for enterprises that need to monitor and control equipment and facilities and integrate with their ERP and CRM back office systems.

Definitions of Web 3.0 vary greatly. Many believe that its most important features are Semantic Web and personalization; some argued that Web 3.0 is where the *computer* is generating new information rather than the human.

The term Semantic Web was coined by Tim Berners-Lee, the inventor of the World Wide Web. He defines the Semantic Web as "a web of data that can be processed directly and indirectly by machines." Humans are capable of using the web to carry out tasks such as reserving a library book or searching

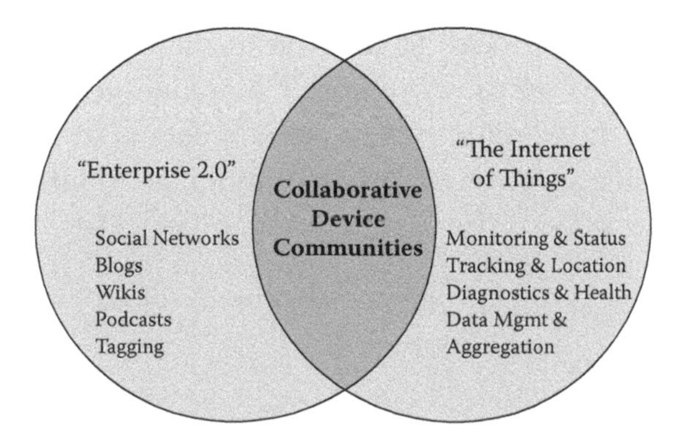

Figure 1.8 Blending of IoT and Enterprise 2.0.

for a low price for a DVD. However, machines cannot accomplish all of these tasks without human direction, because web pages are designed to be read by people, not machines. The Semantic Web is a vision of information that can be readily interpreted by machines, so machines can perform more of the tedious work involved in finding, combining, and acting upon information on the Web.

Some consider the Semantic Web an unrealizable abstraction and see Web 3.0 as the return of experts and authorities to the Web. I share the same thought. If there is no tangible difference but only a conceptual one, the concept of Semantic Web–based Web 3.0 doesn't stand on solid ground. Rather, the Web 3.0 of machine-generated data is more practical, makes more sense, and is possible to implement.

While Web 3.0 arguments are not yet settled, some people have started talking about Web 4.0 [30], the ubiquitous Web.

A fundamental difference between the Internet of People (Web 1.0 and Web 2.0) and the Internet of Things is that in the former, data are generated by people (keyed in by hand, photographed by hand, etc.); in the latter, data are generated by machines, not humans. This difference makes it enough to start a new version of the World Wide Web, or Web 3.0. The data are generated by things and consumed by people and machines via SaaS or XaaS (Everything as a Service), and this model constitutes the basis of Web 3.0 as depicted in Figure 1.9 [74]. We choose to use the term Web 3.0 instead of Web 4.0 based on the concept of machine-generated data in addition to the Semantic Web, which seems to not have much substance up to now. It is too much of a jump to go to Web 4.0.

1.5 Summary

After decades of fast-paced development, telecom networks worldwide now basically satisfy the need for man-to-man

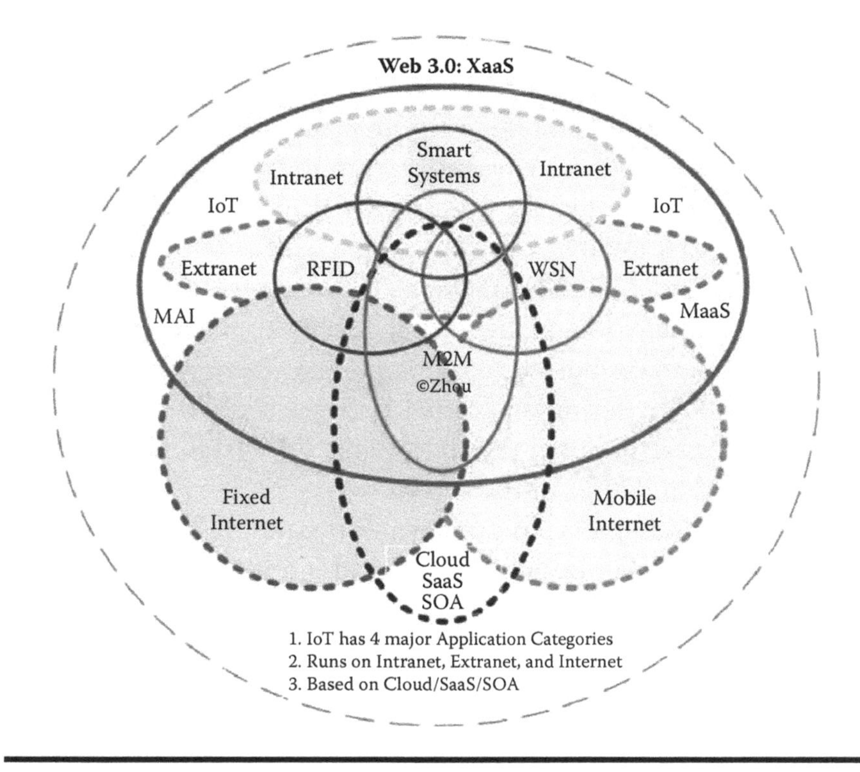

Figure 1.9 Web 3.0: The Internet of Things.

communication anywhere and at any time. However, new demand has arisen for machine-to-machine and machine-to-man, or the Internet of Things, communications. The development of these M2M technologies has attracted greater attention in recent times in light of the "smart Earth" and "Sensing China" concepts proposed by the American and Chinese governments and other parts of the world such as the European Union following the global financial crisis. According to Forrester Research, by 2020 machine-to-machine data exchange will be 30 times greater than the number of exchanges between people. M2M or IoT is therefore considered the next trillion-dollar segment of the international telecom market.

The physical world itself is becoming a connected information system. In the world of the Internet of Things, sensors and actuators embedded in physical objects are linked through wired and wireless networks that connect the Internet. These information systems churn out huge volumes of data that flow to computers for analysis. When objects can both sense the environment and communicate, they become tools for understanding the complexity of the real world and responding to it swiftly.

The Internet of Things and related concepts, terms, and phrases and their potentially vast scope of applications as well as their impacts on business and social life were described in this chapter. The definitions of IoT were described and the author also gave his own definition and understanding, which will be the foundation of the book.

In the next chapter, a more detailed, panoramic view of IoT applications will be introduced and a few concrete vertical applications will be described in greater detail.

Chapter 2

Ubiquitous IoT Applications

2.1 A Panoramic View of IoT Applications

We talked about the big picture and megatrends of the Internet of Things (IoT) in the first chapter, and now we are going to describe the vastly large number of IoT applications and related technologies in a variety of fields in greater detail.

Telemetry is an "ancient" technology that allows remote measurement and reporting of information. Although the term commonly refers to wireless data transfer mechanisms, it also encompasses data transferred over other wired media. Telemetry is synonymous with IoT to some, and it can be regarded as one of the earliest IoT applications. It is closely related to and intertwined with other IoT technologies and applications such as machine-to-machine (M2M) and supervisory control and data acquisition (SCADA). One of the first telemetry applications was developed in 1845 between the Russian czar's Winter Palace and the army's headquarters. In 1874, French engineers built a system of weather and snow-depth sensors on Mont Blanc that transmitted real-time information to Paris. Telecommand and telematics (telecommunication + informatics) were more related to telemetry in earlier times. However, telematics nowadays often refers to vehicle tracking, especially passenger car tracking and global positioning system (GPS) services.

Most recently, the IoT is increasingly finding its way into mainstream news. Executives of large companies and even government officials, such as President Obama and the Chinese premier, are speaking about the possibilities and opportunities of having ubiquitous sensors connected to the Internet.

"The next big revolution that will happen is the Internet of Things," said Cisco chief technology officer Padma Warrior. Although the widespread adoption of IoT will take time, the time line is advancing thanks to improvements in underlying technologies. Advances in networking technologies and the standardization [31] of communication protocols, XML-based data representations, and middleware architectures make it possible to collect data from sensors and devices almost anywhere at any time. Ever-smaller silicon chips are gaining new capabilities, while costs are falling. Massive increases in storage and computing power, available via cloud computing, make number crunching possible at a very large scale and at declining cost. It's easy to speculate on possibilities:

- Radio-frequency identification (RFID) tags that know where your luggage is
- Mesh networks of sensors that can more reliably monitor the changing concentrations of volcanic ash
- Heating, ventilating, and air-conditioning (HVAC) units that can coordinate to act in concert, rather than independently
- Smart sticking plasters that detect microscopic changes in skin condition or blood flow
- An in-vehicle terminal or called an edge device that can detect if you are too sleepy to drive safely

- Surveillance systems that can analyze what they are filming, being alert for security abnormalities
- Smart glasses for the visually impaired that can interpret what you're looking at
- A toothbrush that can let you know if you're not putting enough effort into cleaning the inner sides of your lower right molars
- And all of these devices connected together ...

The arrival of the IoT concept and its worldwide attention is closely relevant to environmental, societal, and economic challenges such as climate change, environment protection, energy saving, and globalization. For these reasons the IoT is increasingly used in a large number of sectors. Key sectors in this context are transportation, healthcare, energy and environment, safety and security, logistics, and manufacturing. M2M and embedded mobile devices are sending mobile data to servers that are increasingly useful and valuable to ERPs [34].

Harbor Research segments the IoT/M2M market into 10 key sectors [32], 30+ subsectors, and countless systems and devices:

- Buildings: Institutional/Commercial/Industrial/Home. HVAC, fire and safety, security, elevators, access control systems, lighting
- Energy and Power: Supply/Alternatives/Demand. Turbines, generators, meters, substations, switches
- Industrial: Process Industries/Forming/Converting/Discrete Assembly/Distribution/Supply Chain. Pumps, valves, vessels, tanks, automation and control equipment, capital equipment, pipelines
- Healthcare: Care/Personal/Research. Medical devices, imaging, diagnostics, monitor, surgical equipment
- Retail: Stores/Hospitality/Services. Point-of sale terminals, vending machines, RFID tags, scanners and registers, lighting and refrigeration systems

- Security and Infrastructure: Homeland Security/Emergency Services/National and Regional Defense. GPS systems, radar systems, environmental sensors, vehicles, weaponry, fencing
- Transportation: On-Road Vehicles/Off-Road Vehicles/ Nonvehicular/Transport Infrastructure. Commercial vehicles, airplanes, trains, ships, signage, tolls, RF tags, parking meters, surveillance cameras, tracking systems
- Information Technology and Network Infrastructure: Enterprise/Data Centers. Switches, servers, storage
- Resources: Agriculture/Mining/Oil/Gas/Water. Mining equipment, drilling equipment, pipelines, agricultural equipment
- Consumer/Professional: Appliances/White Goods/Office Equipment/Home Electronics. M2M devices, gadgets, smartphones, tablet PCs, home gateways

Machina Research classified the IoT/M2M market into 3 categories and 11 segments [35]:

- Intelligent Environment: Intelligent buildings/smart cities and transportation
- Intelligent Living: Automotive/consumer electronics
- Intelligent Enterprise: Health/utilities/manufacturing/ retail and leisure/construction/agriculture and extraction/ emergency services and national security

Per the IoT definition of the previous chapter, the goal of IoT is to achieve pervasive M2M connectivity and grand integration and to provide secure, fast, and personalized functionalities and services such as monitoring, sensing, tracking, locating, alerting, scheduling, controlling, protecting, logging, auditing, planning, maintenance, upgrading, data mining, trending, reporting, decision support, dashboard, back office applications, and others. Those functionalities are common features of IoT systems supported by a common three-tier IoT system architecture that will be described in the latter part of the book.

Beecham Research tracks nine key industries and their associated devices using all principle technologies for connecting them [33]. Such devices range from air-conditioning, access control, and lifts and escalators in the buildings sector to wind turbines, utility meters, and pipelines in the energy/power sector and to closed-circuit television and lone worker solutions in the security/environment sector; from magnetic resonance imaging (MRI) scanners, x-ray machines, and blood analyzers in the healthcare/life sciences sector to telematics systems for cars, trucks, containers, and off-road vehicles and road toll schemes in the transportation sector.

A panoramic view of the IoT applications is shown in Figure 2.1 based on summarizing most of the previously

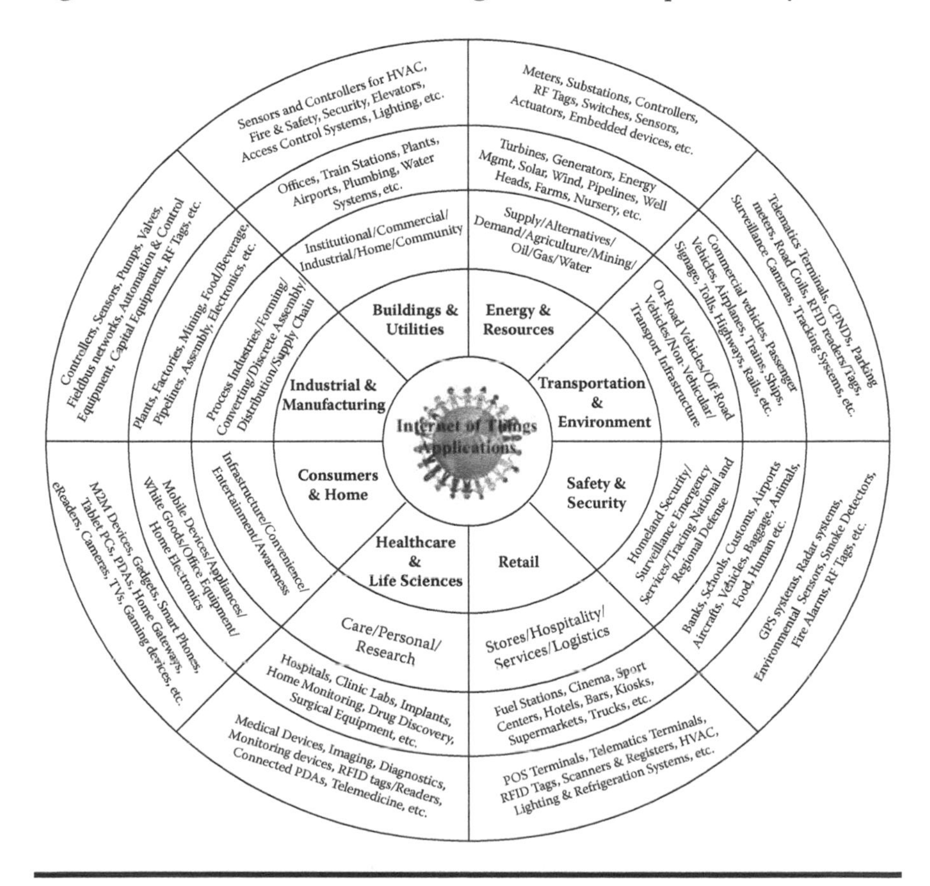

Figure 2.1 A panoramic view of IoT applications.

described industry categories and segments. The first ring is the sectors, the second ring is application groups, the third ring is target objects or sites, and the fourth ring is devices used.

As we see from the previous paragraphs, the term Internet of Things is sometimes used interchangeably with M2M by some market research firms. M2M can be regarded as one of the four sectors under the IoT umbrella; the other ones include RFID, wireless sensor networks (WSN), and SCADA (or called smart systems, industry automation, etc.). Currently, even though almost everyone believes that the IoT market is a huge market, few research reports about the size of the entire IoT market as defined in the last chapter have been produced by market research firms.

Some research firms have reports on two or three of the four IoT sectors, but not all of the four sectors. For example, Harbor Research forecasts that the smart systems [186] and M2M market value will be €280 billion in 2013.

Analysys Mason, a trusted adviser on telecoms, technology, and media, predicts that by 2020, North America will have the most devices per person, with the highest estimate predicting as many as 23.2 devices per person in the region. The Middle East and Africa are expected to have the fewest devices, where estimates are as low as 0.2 per person. The total IoT devices deployed in 2020 will reach 16 billion, a relatively conservative number compared with other predictions. Despite the forecasts for aggressive growth, the IoT has yet to become a mass-market proposition. The IoT still needs to be pulled together into a cohesive and user-friendly package, while security issues also need to be resolved.

Those are predictions tagged with IoT but not necessarily the entire IoT market. However, there are market research reports on the four subsectors of IoT. Several individual market reports will be covered in the next chapter.

Having seen the great potential of the IoT market, many vendors, old and new, have joined forces in this market. A map of comprehensive clusters of IoT vendors based on their focus and position can be found in a publicly available Harbor Research [37] report. The *Connected World Magazine* has published an M2M Top 100 list [188] every year since 2004 [39]. The name of the magazine was changed from *M2M Magazine* [187], which indicates a paradigm change from M2M to a broader IoT coverage. Also, there is a Top 10 list of IoT development for the last two years [190]. Hewlett Packard's Central Nervous System for the Earth was number one on the list in 2010.

2.2 Important Vertical IoT Applications

Before describing the common horizontal technologies underpinning the Internet of Things, we are going to describe some of the important and representative IoT applications in more detail as examples to give the reader more insight and to demonstrate the power and capabilities of IoT technologies or ideologies.

2.2.1 Telematics and Intelligent Transport Systems

Telematics and intelligent transport systems (ITS) are closely related. The IoT technologies and ideologies can be used in telematics as well as ITS, especially in promoting their seamless integration. Telematics and ITS have been a kind of IoT application for a long time. The combined application is called *automobile IoT* in China. It was reported that "the Automotive Mobile Internet of Things has been set as a major project among all the important national projects. At present the relevant materials have been submitted to the State Council. The first batch of funds may total up to ten billion Yuan. By the year of 2020, the amount of controllable (connected) vehicles will reach up to 200 million units." Figure 2.2 shows the scope of China's automobile IoT, which is different from Vehicular Networks [273].

Telematics can be categorized as a subsector of LBS (location-based service; a list of traditional technology-based

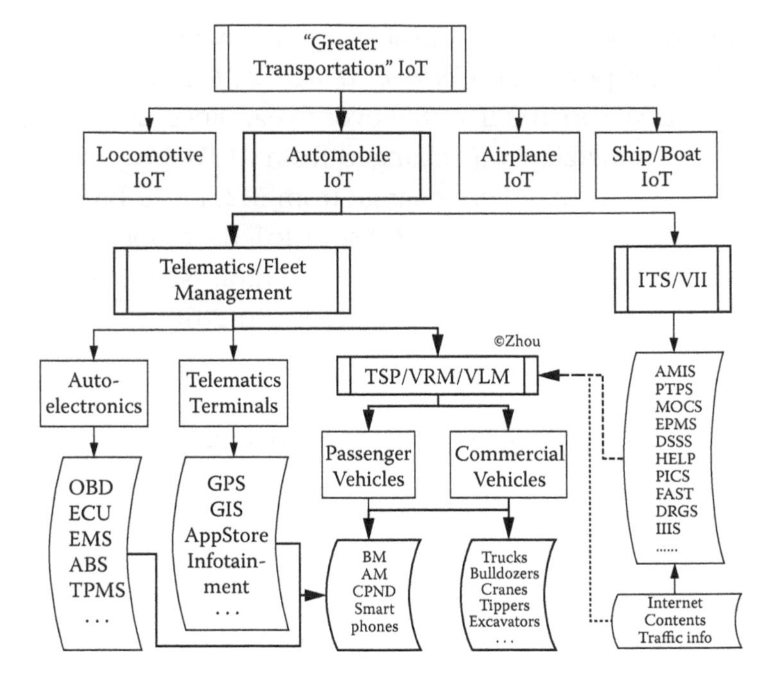

Figure 2.2 Telematics/fleet management/ITS and IoT.

players can be found at http://etutorials.org/Mobile+devices/mobile+wireless+design/Part+Four+Beyond+Enterprise+Data/Chapter+17+Location-Based+Services/LBS+Vendors/; LBS has also been part of social networking services recently with players such as FourSquare and locationary.com). Telematics, as determined by its name, is any integrated use of telecommunication and informatics (Figure 2.3). Its application is within any of the following:

- The technology of sending, receiving, and storing information via telecommunications devices in conjunction with effecting control on remote objects, especially for application in vehicles and with control of vehicles on the move
- GPS technology integrated with computers and mobile communication technology in automotive navigation systems
- The use of such systems within road vehicles, including commercial and (particularly) passenger vehicles

Figure 2.3 Telematics terminal.

The development of auto-electronics as well as telematics has driven the automobile industry into a so-called third-wave automotive industrial revolution. The first automobile revolution was about power, using a high-compression-ratio engine. The second automobile revolution was about control, using microelectronic devices for electronic fuel injection, cruise control, and emission control. And the third revolution is about connectivity (just like M2M) based on telematics for navigation, Internet, ITS integration, and so forth (Figure 2.4).

As of 2010, the cost for vehicle electronics is as high as 40 to 50 percent of the total cost for some vehicles. This is up from 20 percent less than a decade ago [41]. In some luxury cars, the number of microprocessors has reached 50, connected with hundreds of sensors. The sensors and actuators in the vehicles for the monitoring and control of critical units such as the brakes, battery, door locks, safety and security systems, audio/video systems, remote vehicle control, navigation, diagnostic and emission control systems, and others are connected with standard-based buses such as CanBus, LIN, FlexRay, and MOST to the electronic control unit. Types of sensors and actuators in vehicles include sensors and controllers for crash avoidance such as adaptive cruise control

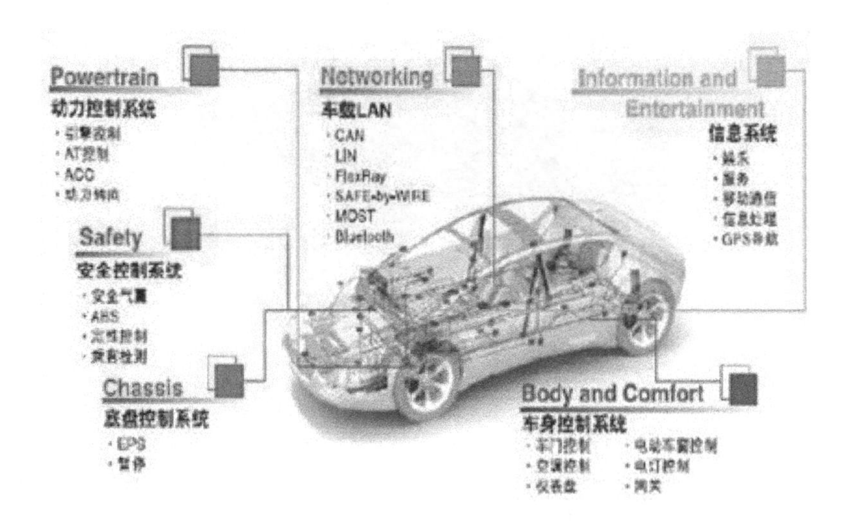

Figure 2.4 In-vehicle networking.

radar, convenience such as remote keyless entry, comfort such as HVAC control, engine sensors such as inlet manifold pressure controller, hybrid and fuel cell such as hydrogen leak detection sensors, vehicle control such as latitude/longitude acceleration controllers, and safety and security such as tire pressure monitoring.

Estimates indicate that the total number of cars owned around the world will reach 1.5 billion in 2020, excluding commercial vehicles and engineering equipment, which account for about one third of the number of cars, making the total automobile number to be around 2 billion in 2020. As the price of telematics terminals keeps going down, it can be expected that telematics terminals with GPS and infotainment capabilities will be a standard device in vehicles just as the radio and CD player are today. This is an enormously huge market.

Auto-electronics exist within a vehicle. Telematics, as a typical M2M application, connects many vehicles to a central server to form a connected vehicle system that provides many services. Organizations providing such services are often called telematics service providers (TSPs). Some of the

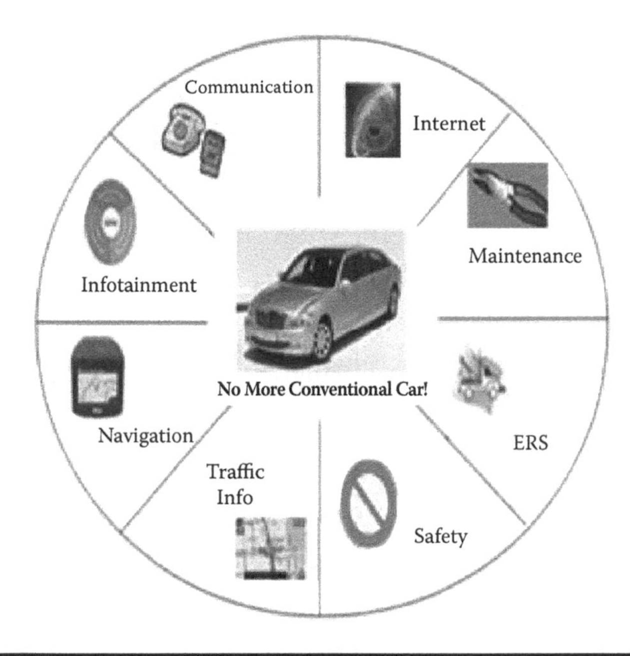

Figure 2.5 Telematics functions/services.

functionalities and services provided by a TSP are shown in Figure 2.5.

NGTP (Next Generation Telematics Pattern, http://www.ngtp.org/) is an open protocol and standard for telematics system architecture created by BMW, Connexis, and WirelessCar. The components of TSP are described in NGTP 1.0 and 2.0 (even though the name is changed to SI in 2.0). Table 2.1 shows a list of major telematics brands.

The telematics terminals can be categorized into BM (before market), AM (after market), and PND (portable navigation device) units. The BM units come with the original vehicle manufacturer and the AM units are integrated into the vehicles later as requested by the vehicle owner. As an indicator of the market size, ABI Research estimates that PND shipments will number more than 150 million units in 2013. However, as more and more telematics device become standard equipment

Table 2.1 Telematics Brands Worldwide

Regions	*Regional Characteristics*	*Telematics Brands*	*Manufacturer Ownership*
USA	1. Vast land 2. Four wireless communications systems coexist	OnStar	GM
		SYNC	Ford
		Connexis	Ygomi independent
		Hughes Telematics	Independent
		Airbiquity	Independent
EU	1. Multilanguage support 2. GSM majority	Tegaron	Daimler Chrysler
		Targa	Fiat Auto
		ATX	Daimler&BMW
		Wireless Car	Volvo Independent
Japan	1. High population density	Internavi	Honda
		CARWINGS	Nissan
		G-BOOK ALPHA	Toyota

from the original vehicle manufacturers, that is, the AM market share is increasing, the PND market has been declining [195].

Figure 2.6 shows a typical architecture of a telematics terminal. A general-purpose embedded middleware layer is often constructed to simplify the development of various and ever-changing applications. Java technology and the universal OSGi middleware framework are often used together to build the embedded middleware.

GENIVI (http://www.genivi.org/) is a nonprofit industry alliance committed to driving the broad adoption of an in-vehicle infotainment (IVI) reference platform. The GENIVI platform—a common software architecture that is scalable

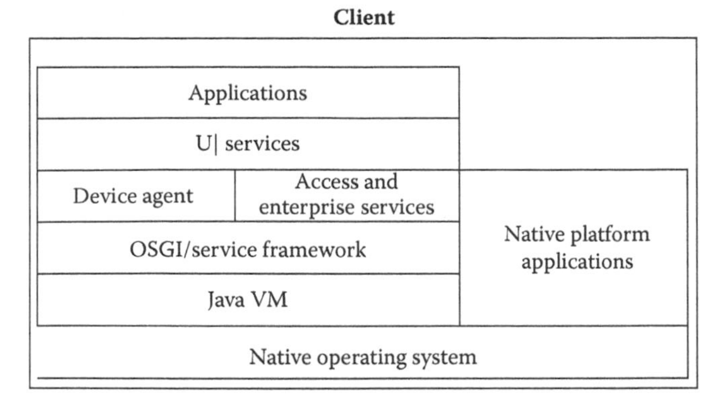

Figure 2.6 Telematics terminal architecture. (From Paolo Bellavista and Antonio Corradi (eds.), *The Handbook of Mobile Middleware,* New York: Auerbach Publications, 2006.)

across product lines and generations—will accelerate the pace at which new and compelling automotive applications are developed and allow new business models to emerge in the in-vehicle infotainment market. It consists of Linux-based core services, middleware, and open application layer interfaces and establishes a foundation upon which automobile manufacturers and their suppliers can add their differentiated products and services.

However, as the iOS and Android application store model become popular, more and more terminals are built on top of Android. The smartphone, the PDA/PC, and the telematics terminal could converge into one screen in the future.

Based on the well-known Gartner hype cycle graph (http://www.gartner.com/technology/research/methodologies/hypecycle.jsp), telematics has passed the hype cycle that happened around 2001, beginning in 1997 when General Motors launched OnStar. The revenue of OnStar surpassed the $1 billion mark in 2010. It is believed that OnStar is the only business unit of GM that didn't lose money from 2005 to 2010. The telematics industry is now on track with healthy and steady developments.

Table 2.2 Fleet Management Brands and Vehicle Manufacturers

Products & Services	*Owners & Providers*
Daimler FleetBoard	Daimler FleetBoard GmbH
Ford CrewChief, Tool Link	Ford Motors
Dynafleet, CareTrack	Volvo (WirelessCar)
JDLink	John Deere
ProductLink, RAC, VIMS	Caterpillar
AWARE Vehicle Intelligence	Navistar (Electronics)
Blue&Me Fleet	Fiat Iveco
Scania Fleet Management	Scania
Squarell Fleet Management	DAF, MAN (third party)
TeloGis, FleetMatics, CFA	Independent third parties

Fleet management, especially GPS-based fleet tracking, is thought by some people as a subsector of telematics known as fleet telematics. However, in some refined market reports, fleet management is regarded as a separate market. The iSuppli corporation market research report lists "Vehicle Tracking and Fleet Management" and "Automotive Telematics" as two markets, with the size of the former market slightly bigger than the latter. ABI Research estimates that the fleet management market is expected to have more than 35 million service connections worldwide in 2013.

Fleet management is for commercial vehicles what telematics is for passenger vehicles. Table 2.2 lists the major truck and engineering equipment manufacturers and their fleet management products and services developed in-house or provided by third parties.

Fleet management (and also telematics) is a subsector of MRM (mobile resource management), which is itself a subsector of the M2M business. According to a 2009 report of C.J. Driscoll & Associates—

- More than 225,000 companies used MRM systems and services at the end of year 2008 in the United States.
- An estimated 3.6 million units are in service with a $1.8 billion market, 75 percent of that from services and software.
- The total U.S. MRM market is projected to grow to 6.5 million units in service by the end of 2012.
- However, the addressable market estimates about 106.6 million units as of 2009, with a lot of room for growth.

As part of MRM, mobile workers are one of the largest segments in the workforce. Any business that fields a sizable mobile workforce faces tough management challenges, including locating and communicating with mobile workers on demand, strengthening dispatching and scheduling capabilities, improving customer quality of experience, and cutting field asset costs and risks. Beyond these challenges, companies are looking to empower their mobile workforces and create additional revenue streams by providing mobile workers with access to back-office applications like enterprise resource planning (ERP) and customer relationship management (CRM) systems.

According to Driscoll, the largest MRM supplier was Qualcomm, with 490,000 units in service by the end of 2009; the second largest was Trimble @Road, which has 250,000 units deployed. In 2011, the author had a meeting with executives from TeloGis, who claim that their TSP services cover 500,000 vehicles. The author led a team and developed a fleet management system called e-Logistics (not NGTP compliant) on top of the general-purpose ^{ez}M2M middleware platform product in 2007 (Figure 2.7). This system has been in operation with China Mobile providing TSP services for nationwide logistics fleet services firms since 2007, currently with 60,000 vehicles from 500+ companies. An M2M service that locates senior people and students was also developed with China Mobile in 2010.

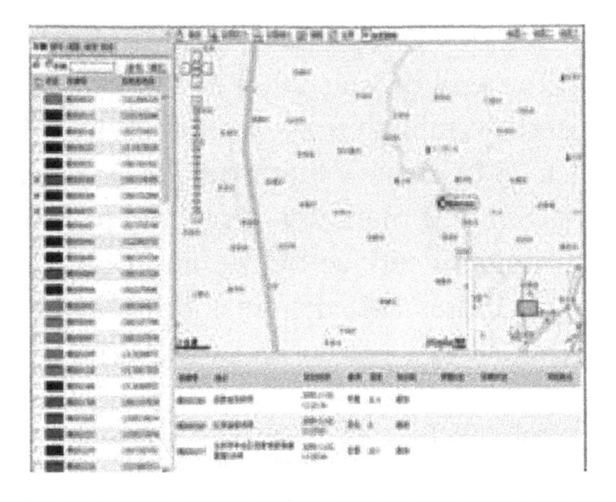

Figure 2.7 e-Logistics user interface.

Telematics and fleet management–based applications can be extended to enable many innovative capabilities:

- Vehicle relationship management has been designed to utilize a vehicle's telematics hardware to provide cost reductions, business efficiencies, and enhanced customer service for automobile manufacturers and their affiliated automobile dealerships.
- Interest has increased across the globe in the benefits of usage-based car insurance, also known as PAYD (Pay as You Drive), which enables vehicle owners to pay reduced car insurance premiums based only on the distances that they drive and the way that they drive.
- Vehicle lifecycle management solution aims to improve customer service, optimize operational processes, lower costs, increase vehicle safety, and improve productivity throughout the automotive design process and supply chain, as well as provides telematics services to vehicle consumers, automotive retailers, car companies, and their suppliers.

The term *intelligent transport systems* (ITS) refers to information and communication technologies (ICT) applied to transport infrastructure and vehicles that improve transport such as transport safety, transport productivity, travel reliability, informed travel choices, social equity, environmental performance, and network operation resilience.

Recent governmental activity in the area of ITS, specifically in the United States, is further motivated by an increasing focus on homeland security. Many of the proposed ITS systems also involve surveillance of the roadways, which is a priority of homeland security. Funding of many systems comes either directly through homeland security organizations or with their approval. Further, ITS can play a role in the rapid mass evacuation of people in urban centers after large casualty events such as a result of a natural disaster or threat. Much of the infrastructure and planning involved with ITS parallels the need for homeland security systems.

According to the U.S. Department of Transportation (DOT) [43], linking vehicles and the transportation infrastructure into an integrated, nationwide system as shown below has been its vision for almost two decades. The VII (vehicle-infrastructure integration) vision, technologies, network, and services are designed to support applications facilitating three major goals: safety, mobility, and e-commerce.

The same vision is shared in Japan, with the goal to reduce the number of vehicle accident fatalities to fewer than 5,000 in 2012, and in the European Union (EU), whose goal was to cut the number of road fatalities by 50 percent in three years. The Next Generation Traffic Management System (UTMS'21) is a new initiative developed by the Universal Traffic Management Society of Japan [44]. In 2003, to realize the original U.S. DOT VII vision, it was determined that the 5.9 GHz dedicated short-range communications (DSRC) would be used by all vehicles by the 2012–2015 time frame. DSRC has been a standard technology used by U.S., EU, and Japanese ITS initiatives (as shown in Figure 2.8). Many other countries are expected to follow.

Figure 2.8 DSRC-based systems.

Apart from DSRC and the aforementioned NGTP and GENIVI, many other alliances and standards organizations are proposing telematics/ITS standards, such as

- Automotive Open System Architecture (AUTOSAR)
- Society of Automotive Engineers (SAE)
- Automotive Multimedia Interface Collaboration (AMI-C)
- 3GPP
- Telecommunications Industry Association (TIA)
- Automatic Terminal Information Service (ATIS)
- Communications for Coordinated Assistance and Response to Emergencies (COMCARE)
- National Emergency Number Association (NENA)
- ISO
- IEEE
- Open Services Gateway Initiative (OSGi)
- ITU
- ESTI

Most of the standards are about the integrated Telematics and ITS systems and applications. Some of the notable ones are OSGi VEG, AUTOSAR, and SAE J2735. Those standards can be employed to work with the DSRC communication standard to realize the VII vision.

2.2.2 Smart Grid and Electric Vehicles

The power grid has evolved into a blended electricity supply and ICT systems as shown in Figure 2.9.

Based on the blending trend, the EPRI (Electric Power Research Institute), an independent nonprofit organization in the United States, proposed the Complex Interactive Networks/Systems Initiative [46], which brought the fundamentals of smart grid together in 1998 as shown in Figure 2.10.

Power SCADA, a technology of IoT characteristics, has long been a stalwart of electric utility operations, becoming increasingly complex as new technologies arrive and new issues emerge on the road to a modern electric smart grid [174]. SCADA/EMS/GMS (energy management system [172], generation management system) supervises, controls, optimizes and

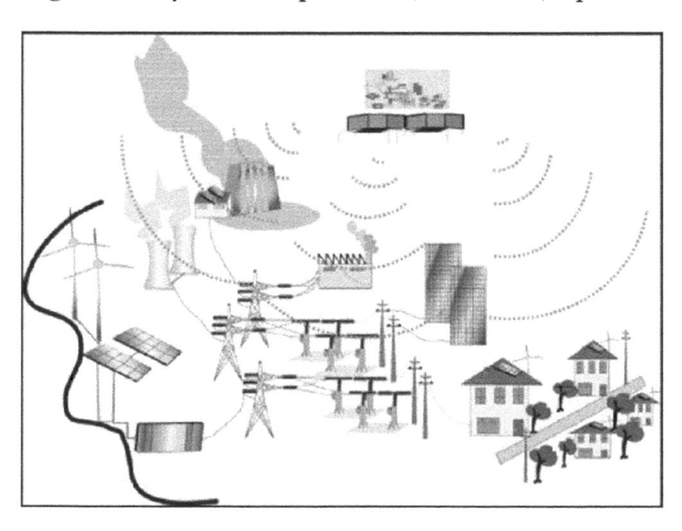

Figure 2.9 The smart grid. (From Melike Erol-Kantarci and Hussein T. Mouftah, "Pervasive Energy Management for the Smart Grid: Towards a Low Carbon Economy," in Hussein T. Shah, Syed Ijlal Ali Ilyas, and Mohammad Mouftah (eds.), *Pervasive Communications Handbook*, Boca Raton, FL: CRC Press, 2011.)

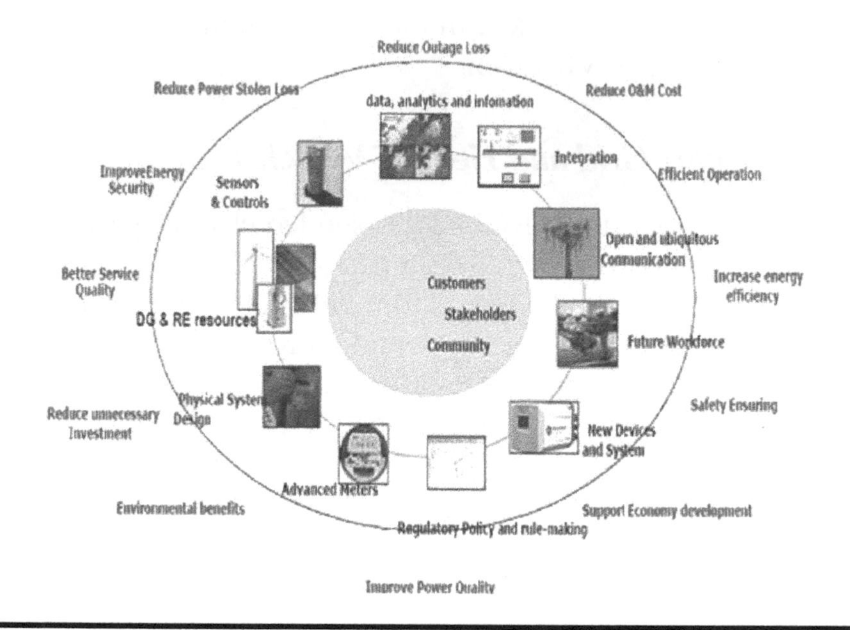

Figure 2.10 Smart grid value chain and stakeholders.

manages generation and transmission systems. SCADA/DMS (distribution management system) performs the same functions for power distribution networks. Both systems enable utilities to collect, store, and analyze data from hundreds of thousands of data points in national or regional networks, perform network modeling, simulate power operation, pinpoint faults, preempt outages, and participate in energy trading markets. These systems are a vital part of modern power networks and are enabling the development of smart grids.

In 2000, EPRI proposed the IntelliGrid initiative. The U.S. Department of Energy (DOE) created the GridWise [193] program during the same time frame. In 2003, the U.S. DOE published its "Grid 2030" report [47]: "'Grid 2030' energizes a competitive North American marketplace for electricity. It connects everyone to abundant, affordable, clean, efficient, and reliable electric power anytime, anywhere. It provides the best and most secure electric services available in the world." In 2005, the European Technology Platform SmartGrids was set up to create a joint vision for the European networks of 2020 and beyond, and the term *smart grid* became widely used since then. The Modern Grid Initiative [175] was created by the National Energy Technology Laboratory for the U.S. DOE in 2007. A NIST (National Institute of Standards and Technology) smart grid interoperability standard [176] specification was proposed in 2010 by Gary Locke, the U.S. secretary of commerce, and Patrick D. Gallagher, director of NIST.

Smart grid technologies have emerged from earlier attempts at controlling, metering, and monitoring. In the 1980s, automatic meter reading was used for monitoring loads from large customers and evolved into the AMI (advanced metering infrastructure) of the 1990s, with meters that could store how electricity was used at different times of the day. Smart meters add continuous communications so that monitoring can be done in real time and can be used as a gateway to demand response-aware devices and "smart sockets" in the home.

Monitoring and synchronization of wide area networks were revolutionized in the early 1990s when the Bonneville Power Administration expanded its smart grid research with prototype sensors that were capable of very rapid analysis of anomalies in electricity quality over very large geographic areas. The culmination of this work was the first operational wide area measurement system (WAMS) in 2000. The IoT technologies and ideologies play an important role in this approach. Other countries are rapidly integrating this technology. For example, China will have a comprehensive national WAMS system when its current five-year economic plan is complete in 2012, with plans to have PMU (phasor measurement unit) sensors at all generators of 300 or more megawatts and all substations of 500 or more kilovolts. In 2009, China announced an aggressive framework for "strong grid" [166] deployment. Compared with that in the United States and in Europe, China's smart grid appears to be more transmission centric [48].

A number of challenges face the power industry that its communications infrastructure is not currently prepared to

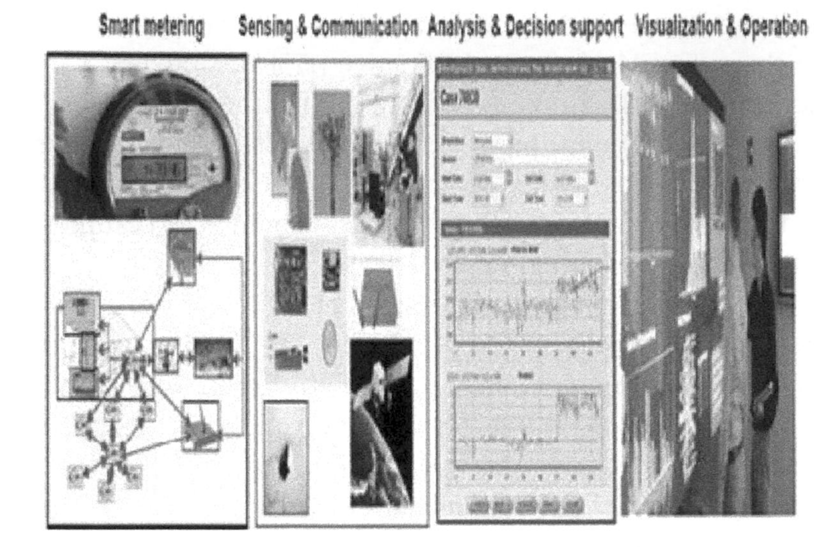

Figure 2.11 Smart grid technologies.

address. A power system making use of an integrated electrical and communications systems architecture (as shown in Figure 2.11) should be as follows:

- Self-healing and adaptive, applying automated applications for protection, fault detection, fault location, sectionalization, and automatic service restoration over wide areas of the service territory
- Interactive with consumers and markets, permitting real-time pricing, energy trading, and load management
- Optimized to make the best use of aging equipment, personnel from multiple organizations, and other resources in a competitive environment
- Predictive, scheduling maintenance ahead of time to prevent rather than just react to emergencies
- Distributed, permitting activities such as generation, metering, load shedding, and others to be easily performed at different locations and by different organizations

- Integrated, merging the previously separate functions of monitoring, control, protection, maintenance, energy management, distribution management, business, and corporate information technology
- Secure, protecting vital infrastructure from cyber or physical attack

Although these functions are performed today by various utilities, there is much variation in the level of implementation, and they are generally not performed on a wide enough scale to address the level of problems faced by the grid today. This scenario is exactly the same as other vertical sectors of the IoT landscape.

The Integrated Energy and Communications System Architecture IntelliGrid Architecture project of Consortium for Electric Infrastructure to Support a Digital Society (CEIDS), Electricity Innovation Institute (E2I), and EPRI has developed an open, standards-based systems architecture for data communications and distributed computing infrastructure that will enable the integration of a wide variety of intelligent electric power system components and transducer devices at a much larger scale and higher levels.

A great many smart grid definitions exist: some functional, some technological, and some benefits oriented. A common element to most definitions is the application of advanced sensor technologies, two-way communications, and distributed processing to the power grid, making data flow and information management central to the smart grid.

The smart grid ecosystem and its drivers and components are described in many research works such as [196,198]. IT giants such as IBM, Cisco, Microsoft, Intel, Oracle, and others have actively participated in many relevant works on smart grids. One of the players of special interest is Google. Google joined the smart grid party by announcing its PowerMeter program in 2009, which aimed to ultimately become an open platform for home energy information. A home energy gadget on the iGoogle home page shows how much energy is being used. The gadget tracks historical data and forecasts future trends. Underneath the PowerMeter gadget is an open systems platform that Google equates to Google Maps, the highly successful geospatial system that has become the foundation for thousands of applications. However, Google shut down the PowerMeter site in 2011 after two years due to lack of users. The reason is not that the PowerMeter services are not needed by users, but rather that it's probably too early in the smart grid development stage, among many other reasons.

Smart grid research will have to consider incorporating renewable energies into the power network and the provisioning of electric vehicles. In a true smart grid, electric cars will not only be able to draw on electricity to run their motors, but they will also be able to do the reverse: send electricity stored in their batteries back into the grid when it is needed. On average, American automobiles get driven for just one hour each day. Most cars are going to have lots of extra battery capacity. Electrifying the entire vehicle fleet would provide more than three times the power generated in the United States. On the other hand, it's important to make sure people are not charging at the very peak time, like late afternoon when the electricity grid is already weighted down by demands like air-conditioning.

Vehicle-to-grid (V2G) describes a system in which plug-in electric vehicles (EVs), such as battery electric vehicles and plug-in hybrid electric vehicles, communicate with the power grid to sell demand response services either by delivering electricity into the grid or by throttling their charging rate. Since most vehicles are parked an average of 95 percent of the time, their batteries could be used to let electricity flow from the car to the power lines and back, with a value to the utilities of up to $4,000 per year per car. V2G is a version of battery-to-grid power applied to vehicles. There are three different versions of the vehicle-to-grid concept:

- A hybrid or fuel cell vehicle, which generates power from storable fuel, uses its generator to produce power for a utility at peak electricity usage times. Here the vehicles serve as a distributed generation system, producing power from conventional fossil fuels or hydrogen.
- A battery-powered or plug-in hybrid vehicle, which uses its excess rechargeable battery capacity to provide power to the electric grid in response to peak load demands. These vehicles can then be recharged during off-peak hours at cheaper rates while helping to absorb excess nighttime generation. Here the vehicles serve as a distributed battery storage system to buffer power.
- A solar vehicle, which uses its excess charging capacity to provide power to the electric grid when the battery is fully charged. Here the vehicle effectively becomes a small renewable energy power station. Such systems have been in use since the 1990s and are routinely used in the case of large vehicles, especially solar-powered boats.

One of the biggest challenges for the mass adoption of electric vehicles by consumers is range anxiety. Having driven traditional cars, where infrastructure for refueling is abundantly established, consumers are still wary of the dead car situation that an EV might pose. Although EV OEMs and battery manufacturers are constantly working to improve battery range, the associated added costs with the increased range pose a threat for mass adoption as well. In the long run, range anxiety might be tempered by producing cost-effective, long-range batteries through constant research and by establishing an adequate public charging smart grid infrastructure. But in the short-term, according to Frost and Sullivan [197], the answer is that OEMs and suppliers should resort to telematics and connected services as a solution to range anxiety

Toyota and Microsoft launched a $12 million venture in 2011 to bring telematics to Toyota's vehicle via the cloud, allowing owners to connect to information services and manage the batteries in their electric vehicles. They will create a global network based on the Windows Azure cloud computing platform, creating a system by which cars like the forthcoming RAV4 EV and plug-in Prius communicate with and draw power from the grid. We'll see it first in the electric and plug-in hybrids Toyota introduced on a limited scale in 2012.

Electric vehicles need to be admired not only for their bodies but also for their brains. According to Pike Research [199], the second wave of EVs and plug-in electric vehicles are likely to be even smarter than the first as automakers are enhancing their telematics features. Toyota is partnering with Microsoft so that its vehicles can communicate with Microsoft's cloud computing technology. Toyota's Media Service division is peering into the home energy management market and will enable its plug-in electric vehicles and their accompanying mobile applications to control electricity consumption in both the car and the home.

In addition to the automakers themselves, telematics companies focusing on EVs abound, including Airbiquity, Automatiks, and Telogis, just to name a few. These companies are extending applications such as green routing to avoid traffic and produce energy efficiency for fleet managers. This increase in the brain power could spell trouble for the makers of electric vehicle charging equipment, who want their devices and not the cars themselves to be the center of smart vehicle charging.

Companies such as GE, Siemens, Ecotality, and Coulomb Technologies see the car–home connection as a great opportunity to expand the value of the equipment and are working on integrating their software with home energy management. They are ahead of the automakers in this regard today, but may not be for long if the major vehicle manufacturers follow Toyota's lead.

The two-way provisioning capabilities of electric vehicles are widely seen as a killer application* to smart grid.

* Killer application or "killer app" is a buzzword that describes a software application that surpasses all of its competitors. (http://www.investopedia.com/terms/k/killerapplication.asp#ixzz22Ogaukm1http://www.investopedia.com/terms/k/killerapplication.asp#axzz22Of3x3Hh)

2.2.3 Smarter Planet and Smart Buildings

Smarter Planet is an IBM initiative mentioned in President Obama's speeches. The initiative seeks to highlight how forward-thinking leaders in business, government, and civil society around the world are capturing the potential of smarter systems to achieve economic growth, near-term efficiency, sustainable development, and societal progress. Many of the challenges the planet faces are concentrated in cities. Cities struggle with traffic congestion, water management, environment protection, public utility management, smart grids, healthcare solutions, building energy efficiency, and rail transportation issues, to name a few. These issues have historically been difficult to manage because of their size and complexity. But with new ways of monitoring, connecting, and analyzing the systems, business, civic, and nongovernmental leaders are developing new ways to address those issues.

According to Forrester, a smart city is one that "uses information and communications technologies to make the critical infrastructure components and services of a city (administration, education, healthcare, public safety, real estate, transportation, utilities, and so on) more aware, interactive and efficient" [274].

Here are some examples for ways that technologies can affect different "systems" required to keep a city up and running, in good health:

- Intelligent sensors that keep tabs on things and places
- Business intelligence and analytics applications that can help slice, dice, and make sense of the data
- Wireless networks and other mobile communications technologies
- Alerts and workflow automation

In building a smart city, ICT has a fundamental role to play. The adoption of hardware, software, and services gives way to the creation of a new, holistic, ICT ecosystem which IDC refers to as "Intelligent X." IDC defines Intelligent X as a technology ecosystem that integrates the following three areas [45]:

- Smart devices involving M2M/telemetry capabilities
- High-speed ubiquitous communications networks
- Intelligent software and services to process, consolidate, and analyze data in order to transform industry-specific business processes

IDC has outlined seven categories of applications for smarter cities [200]: health; home; sports and leisure; education; transport; buildings; and city services, safety, security, and emergency response. IDC also published a more detailed list of "Intelligent X" core competencies for building a smarter city: city strategic planning, government, networks, devices, verticals, marketing, public relations, and finance.

It is not difficult to find out that the technologies used to build a smarter planet or smarter cities as defined by IDC or Forrester is almost the same as those defined in the Internet of Things in the previous chapter. So the Internet of Things plays an important role in building a smarter planet and smarter cities.

In Cisco's blueprint for its Smart + Connected Communities initiative [201], connected and sustainable mobility and connected and sustainable energy are two of the three areas that are most important in building a smart city; some important topics related to these have been described in the last two sections. The third area is connected and sustainable buildings, which will be described in greater detail later in this section. The other two areas are connected and sustainable work and sustainable socioeconomics.

Smart buildings are the building blocks of a smart city, which are building blocks of the smarter planet. An intelligent green building is managed by a building management system (BMS) or an interconnected, integrated, and intelligent BMS. All four IoT technologies—SCADA, M2M, RFID, and WSN—can be used in a BMS. A BMS usually controls and monitors the building's mechanical and electrical equipment that integrates the BAS (building automation system), security and alerting system, fire alarming system, closed-circuit TV video surveillance system, access control system, power and lighting system, elevator, broadcasting and background music system, parking system, network and cable TV management system, PMS (property management system), and even office automation system. With the advent of IoT, more systems such as energy efficiency management (power and water usage metering and submetering) are added into BMS.

Although some people have blurred the difference between a BMS and a BAS, we believe that a BAS should be part of an integrated BMS. A BMS usually uses higher level Internet and wireless mesh network protocols as well as open standards such as DeviceNet, ZigBee, EnOcean (energy harvesting technology), SOAP, and XML, and builds on top of a middleware platform such as a three-tiered Java application server for web-based access anywhere, anytime. An open BMS standard named oBIX (open Building Information eXchange) was proposed and maintained by OASIS.

Figure 2.12 is a BMS system product named ^{ez}IBS developed by the author's team (^{ez}IBS is the BMS market leader in

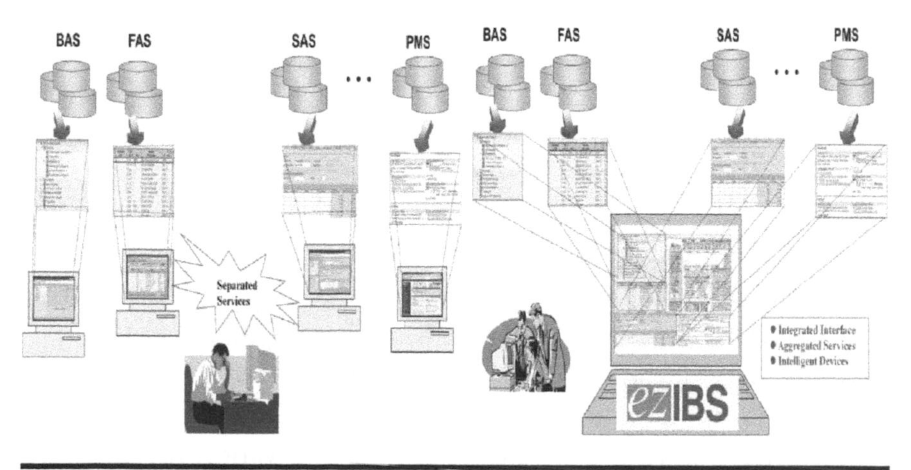

Figure 2.12 Integrated building system.

China and it has been referenced in college textbooks and used as a study system by students [254]). It is a system built on top of a general-purpose three-tier JavaEE middleware platform. With ezIBS, all the component systems (on the left of the figure) in a building are integrated into one interconnected, intelligent system that provides integrated services.

A BMS system is an example of a human machine interface (HMI/SCADA); similar systems include Wonderware and Tridium. BMS evolved from BAS, and BAS evolved from direct digital control and programmable logic controller. Due to the wide adoption of the object linking and embedding for process control standard and historic reasons, most of such systems (such as Wonderware [Intouch, IAS], Rockwell [FTView, RsView], Siemens [WinCC], Axeda [Wizcon], and ArcInformatique [PcVue]) were built using Windows technologies. Newer systems such as Tridium (of Honeywell) and ezIBS are based on open JavaEE middleware technologies.

A BAS is an example of a distributed control system, which, in most cases, covers the HVAC systems of a building, while a BMS is like an information system that does a grand integration of everything in the building (as shown in Figure 2.12). A BAS's core functionality keeps the building climate within a specified range and monitors system performance and device failures. A BAS is usually configured in a hierarchical manner using lower level protocols as CAN-bus, Profibus, BACnet, LonWorks, and Modbus.

ESPC (energy savings performance contract) is an alternative financing mechanism authorized by the U.S. Congress and designed to accelerate investment in cost-effective energy conservation measures in existing federal buildings. ESPCs are regulations created by the Federal Energy Management Program [173] of the U.S. DOE as required by the Energy Policy Act of 1992, which authorizes federal agencies to use private-sector financing to implement energy conservation methods and energy efficiency technologies.

An ESPC is a partnership between a federal agency and an energy service company (ESCO). The ESCO conducts a comprehensive energy audit for the federal facility and identifies improvements to save energy. In consultation with the federal agency, the ESCO designs and constructs a project that meets the agency's needs and arranges the necessary financing. The ESCO guarantees that the improvements will generate energy cost savings sufficient to pay for the project over the term of the contract. ESCOs employ a BEMS (building energy management system) to fulfill ESPCs.

A BEMS is a system that facilitates management and control of building facilities while also realizing energy savings and increasing comfort of building users by making full use of state-of-the-art information technology. A BEMS is similar to a BMS, yet initially focused specifically on energy; while a BMS does have energy management aspects to it, it also includes the monitoring of fire systems and security systems among other building and mechanical controls. BEMS are brought to market by vendors who solely focus on energy management as their means to penetrate customer channels.

According to Pike Research, many BEMS projects are implemented with existing BMS installations. Historically, BMS players such as Honeywell, Johnson Controls, and Siemens have dominated the energy management market for commercial buildings. However, newer, more nimble players, like EnerNOC (convergence) and BuildingIQ (new entrant), are beginning to increase market share and help define a new market.

As a summarization, Figure 2.13 shows the product portfolios of the company the author helped build (the software application systems based on an JavaEE middleware platform) for comprehensive intelligent buildings and smart city applications.

Another IoT application on buildings is the home automation segment. Home automation, also called domotics, is the residential extension of building automation. It is automation of the home, housework, or household activity. Home

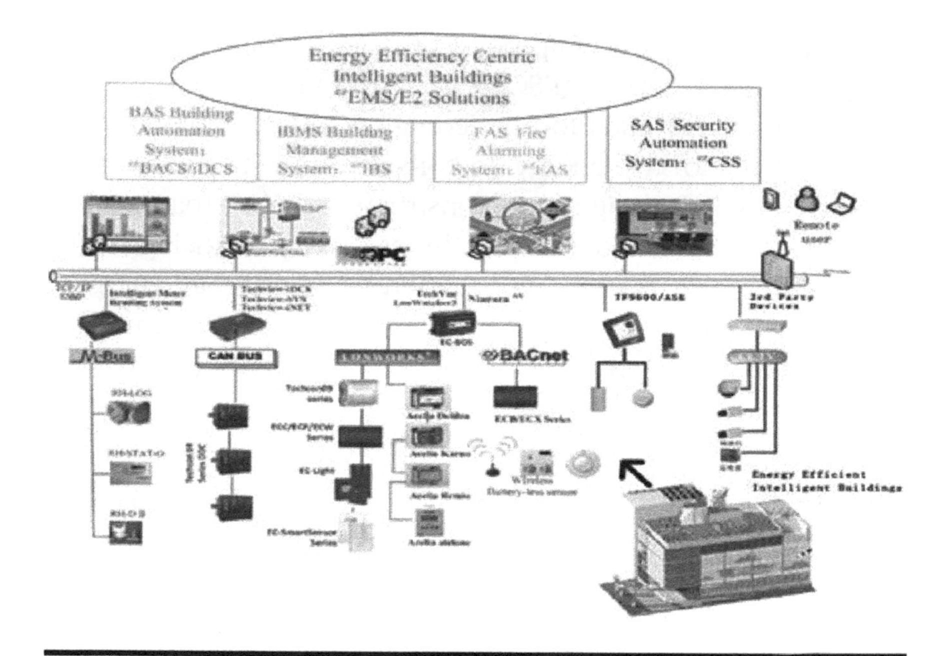

Figure 2.13 IoT applications for buildings.

automation may include centralized control of lighting, HVAC, appliances, and other systems to provide improved convenience, comfort, energy efficiency, and security. Home automation for the elderly and disabled can provide increased quality of life for people who might otherwise require caregivers or institutional care.

2.3 Summary

As an introduction to the widespread, ubiquitous IoT applications, we gave a panoramic view of the IoT application landscape at the beginning of this chapter. We then described three important IoT sectors that IoT technologies and ideologies apply and dominate.

In the next chapter, the four-pillar classification of the Internet of Things will be proposed and outlined. The technologies and applications of each pillar will be described in detail.

Chapter 3

Four Pillars of IoT

3.1 The Horizontal, Verticals, and Four Pillars

Applications of the Internet of Things (IoT) have spread across an enormously large number of industry sectors, and some technologies have been used for decades as described in the previous chapter. The development of the vertical applications in these sectors is unbalanced. It is very important to sort out those vertical applications and identify common underpinning technologies that can be used across the board, so that interconnecting, interrelating, and synergized grand integration and new creative, disruptive applications can be achieved.

One of the common characteristics of the Internet of Things is that objects in a IoT world have to be instrumented (step 3 in Figure 3.1), interconnected (steps 2 and 1), before anything can be intelligently processed and used anywhere, anytime, anyway, and anyhow (steps 1 and 2), which are the 5A and 3I [180] characteristics.

Another common feature that IoT brought to information and communications technology (ICT) systems is a fundamental change in the way information is generated, from

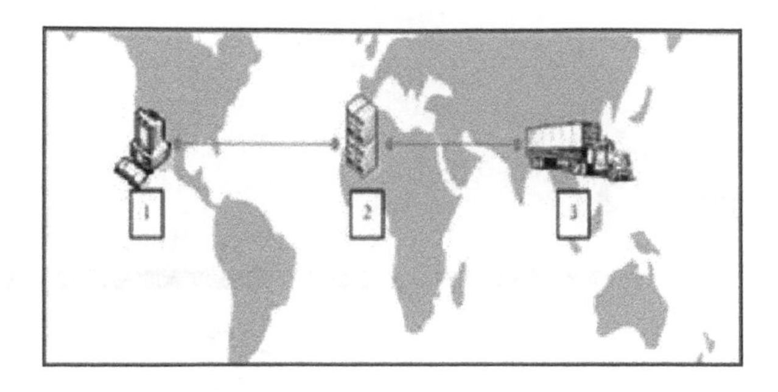

Figure 3.1 3I and 5A.

mostly manual input to massively machine-generated without human intervention.

To achieve such 5A (anything, anywhere, anytime, anyway, anyhow) and 3I (instrumented, interconnected, and intelligent) capabilities, some common, horizontal, general-purpose technologies, standards, and platforms, especially middleware platforms based on common data representations just like the three-tiered application server middleware, HTML language, and HTTP protocol in the Internet/web arena, have to be established to support various vertical applications cost effectively, and new applications can be added to the platform unlimitedly.

Most of the vertical applications of IoT utilize common technologies from the networking level and middleware platform to the application level, such as standard wired and wireless networks, DBMS, security framework, web-based three-tiered middleware, multitenant PaaS (platform as a service), SOA (service-oriented architecture) interfaces, and so on. Those common technologies can be consolidated into a general-purpose, scalable framework and platform to better serve the vertical applications as demonstrated in [202].

Service-management platforms (SMPs) are the key to entry into the machine-to-machine (M2M) market. They allow for the essential connectivity management, intelligent rate-plan management, and customer self-service capability that are today's fundamental prerequisites for providing a successful, managed M2M service. Consequently, with its acquisition of

Telenor Connexion's M2M SMP technology and the staff related to the platform's development, Ericsson has taken a decisive step into the market. Ericsson has built a horizontal platform for the 50 billion M2M market's vertical telematics, medical, utilities, and government applications [203].

Telenor Objects was formed in July 2009 by researchers and developers in Telenor Norway and Telenor R&I. The two entities had individually been working on piloting managed M2M services since 2007, with an RFID (radio-frequency identification) focus in Telenor Norway, and a focus on trace-and-track initiatives in Telenor R&I. Telenor Objects [104] aims to provide a layered and horizontal architecture for connecting devices and applications. The company's platform, dubbed Shepherd, adheres to ETSI's standardization initiative on connected objects and provides a device library as well as a set of enablers to device and application providers. In addition, Shepherd includes a range of operational management services.

As a driver for connecting devices to the Internet of Things, Telenor Objects is a founding member of coosproject.org (Connected Objects Operating System), a general-purpose, modular, pluggable, and distributable open source middleware platform in Java, designed for connecting service and device objects that communicate via messages and enabling monitoring and management. (The targeting devices totaled 2.675 trillion according to Telenor Objects and Harbor Research's Intelligent Device Hierarchy at http://www.harborresearch.com/_literature_32606/News.htm.) The initiative is among several newly established steps by Telenor into the open source and open innovation sphere.

The key benefits of horizontal standard-based platforms will be faster and less costly application development and more highly functional, robust, and secure applications. Similar to the market benefit of third-party apps (e.g., Apple's application store) running on smartphone platforms, M2M applications developed on horizontal [183] platforms will be able to make easier use of underlying technologies and services. Application developers will not have to pull together the entire value chain or have expertise in esoteric skill sets. This will dramatically increase the rate of innovation in the industry in addition to creating more cross-linkages between various M2M applications.

In an issue of the M2M (now *Connected World*) magazine's cover story in 2007 [50], editorial director Peggy Smedley introduced a graphic that encapsulates the ever-expanding M2M landscape. The graphic covers the "six pillars" of M2M technology, representing market segments that involve networking physical assets and integrating machine data into business systems. The six pillars of M2M are as follows:

1. Remote monitoring is a generic term most often representing supervisory control, data acquisition, and automation of industrial assets.
2. RFID is a data-collection technology that uses electronic tags for storing data.
3. A sensor network monitors physical or environmental conditions, with sensor nodes acting cooperatively to form/maintain the network.
4. The term *smart service* refers to the process of networking equipment and monitoring it at a customer's site so that it can be maintained and serviced more effectively.
5. Telematics is the integration of telecommunications and informatics, but most often it refers to tracking, navigation, and entertainment applications in vehicles.
6. Telemetry [185] is usually associated with industrial-, medical-, and wildlife-tracking applications that transmit small amounts of wireless data.

However, there is plenty of overlap among the pillars in this graphic. Pick any application of M2M and chances are it fits into more than one of the six pillars. Take fleet management as an example. It is certainly remote monitoring. It can be considered a smart service depending on who's doing the monitoring. It may have elements of telematics. It fits the technical definition of telemetry. And, there may even be RFID tags or a sensor network onboard.

In this book, a four-pillar graphic is introduced for the broader IoT universe. The four pillars of IoT are M2M, RFID, WSNs (wireless sensor networks), and SCADA (supervisory control and data acquisition):

- M2M uses devices (such as an in-vehicle gadget) to capture events (such as an engine disorder), via a network (mostly cellular wireless networks, sometimes wired or hybrid) connection to a central server (software program), that translates the captured events into meaningful information (alert failure to be fixed).
- RFID uses radio waves to transfer data from an electronic tag attached to an object to a central system through a reader for the purpose of identifying and tracking the object.
- A WSN consists of spatially distributed autonomous sensors to monitor physical or environmental conditions, such as temperature, pressure, motion, or pollutants, and to cooperatively pass their data through the network, mostly short-range wireless mesh networks, sometimes wired or hybrid, to a main location. (Methley et al. [62] reports on the overlaps or coverage differences when WSN was compared with M2M and RFID; SCADA or smart system was not mentioned in the report.)
- SCADA is an autonomous system based on closed-loop control theory or a smart system or a CPS that connects, monitors, and controls equipment via the network (mostly wired short-range networks, a.k.a., field buses, sometimes wireless or hybrid) in a facility such as a plant or a building.

The term *SCADA* was picked as one of the pillars of IoT over the terms *smart system* and *CPS*. CPS [28] is more of an academic term, and EPoSS defines *smart system* as "miniaturized devices that incorporate functions of sensing, actuation, and control" [22]. Both of these can be considered parts of the extended scope of SCADA or ICS (industrial control system) under the IoT umbrella.

Smart systems evolved from microsystems. They combine technologies and components from microsystems (miniaturized electric, mechanical, optical, and fluid devices) with knowledge, technology, and functionality from disciplines like biology, chemistry, nano sciences, and cognitive sciences.

However, Harbor Research [32] defines smart systems as a new generation of systems architecture (hardware, software, network technologies, and managed services) that provides real-time awareness based on inputs from machines, people, video streams, maps, news feeds, sensors, and more that integrate people, processes, and knowledge to enable collective awareness and decision making. Based on this definition, a smart system is close to an industrial automation system, a facility management system, or a building management system.

Harbor Research's definition is close to what a SCADA system covers. Due to the difference of the definitions of Harbor and EPoSS, SCADA is chosen as one of the four pillars.

There is much less overlap between these four pillars compared with those of the six-pillar categorizations of M2M. The clear categories of the four pillars and the distinct networking technologies are shown in Table 3.1 and Figure 3.2.

Table 3.1 Four Pillars of IoT and Their Relevance to Networks

Four Pillars and Networks	*Short-Range Wireless*	*Long-Range Wireless*	*Short-Range Wired*	*Long-Range Wired*
RFID	Yes	Some	No	Some
WSN	Yes	Some	No	Some
M2M	Some	Yes	No	Some
SCADA	Some	Some	Yes	Yes

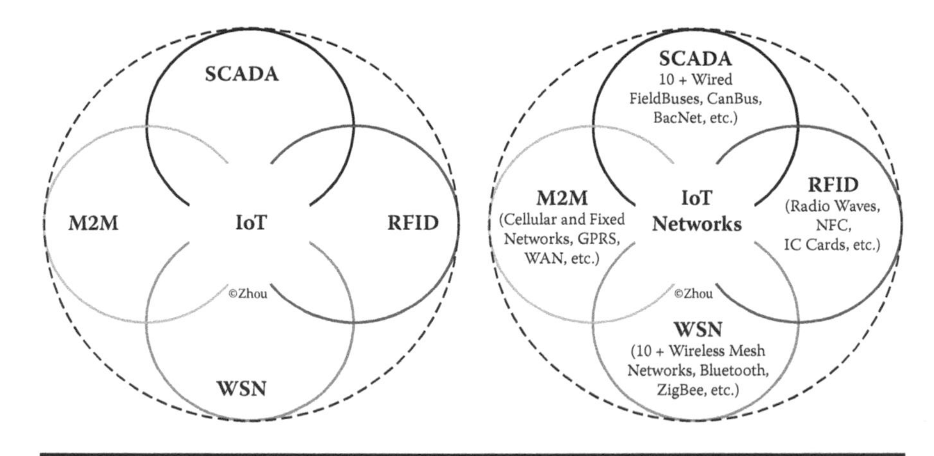

Figure 3.2 The four pillars of IoT paradigms and related networks.

The Strategy Analytics research firm also categorized the IoT networks as wired (stationary) and wireless (mobile), and compared their market value and ease of integration as early as 2004 [204].

IoT is the glue that fastens the four pillars through a common set of best practices, networking methodology, and middleware platform. This enables the user to connect all of their physical assets with a common infrastructure and a consistent methodology for gathering machine data and figuring out what it means. Take away the glue, and end users are left with multiple application platforms and network accounts. The true power of the Internet of Things occurs when it is working behind the scenes (just like Mark Weiser said about ubiquitous computing) and sharing a common platform, which can't happen if companies have to manage multiple, independent systems.

3.2 M2M: The Internet of Devices

Although the rest of the world may not agree, in the United States, *machine-to-machine* is a more popular term than the *Internet of Things*, thanks perhaps to *M2M Magazine*'s efforts since 2004. Two of the six pillars, remote monitoring and smart service, are features or functions of an IoT system rather than pillars. Conceptually, the terms M2M, RFID, and WSN are similar, but when the underlying communication network is taken into consideration, they are quite different segments.

In this book, the term M2M is restricted to refer to device connectivity technologies, products, and services relevant to the cellular wireless networks operated by telco companies. In fact, most of the M2M market research reports assume M2M modules are simply just cellular modules. Table 3.2 showcases the major applications. However, there is overlap between M2M and the consumer electronics applications. The consumer electronics offerings include the following (as opposite to the traditional M2M offerings shown in Table 3.2):

- Personal navigation devices
- eReaders
- Digital picture frames
- People-tracking devices
- Pet-tracking devices
- Home security monitors
- Personal medical devices

ABI Research forecasts that the M2M market is expected to reach more than 85 million connections globally by 2012, and more than 200 million by 2014, with a total market valuation of approximately $57 billion, with utilities (automatic meter reading, telemetry) and automotive (telematics) the clear winners. In fact, it has been assumed that M2M comprises telematics and telemetry [42]. However, Analysys Mason predicts telemetry (utilities, etc.) will outperform telematics in the long run [205].

iSuppli's research depicts the worldwide cellular M2M module market by vertical applications in millions of dollars and the market shares of major vendors [206]. Juniper Research

Table 3.2 Application Areas for Cellular M2M

Industry	*Example Application*	*Benefits*
Medical	Wireless medical device	Remote patient monitoring
Security	Home alarm and surveillance	Real-time remote security and surveillance
Utility	Smart metering	Energy, water, and gas conservation
Manufacturing	Industrial automation	Productivity and cost savings
Automotive	Tracking vehicles	Security against theft
Transport	Traffic systems	Traffic control for efficiency
Advertising and public messaging	Billboard	Remote management of advertising displays
Kiosk	Vending	Remote machine management for efficiency and cost savings
Telematics	Fleet management	Efficiency and cost savings
Payment systems	Mobile transaction terminals	Mobile vending and efficiency
Industrial automation	Over-the-air diagnosis and upgrades	Remote device management for time savings and reduced costs

estimates there will be approximately 412 million M2M mobile connected devices in the marketplace by 2014 [207].

The number of cellular M2M devices surpassed the number of mobile phones for the first time in Europe in 2010, just a few months later than the time predicted by e-Principles in 2003.

According to Beecham Research in August 2011, Cisco recently announced dedicated routers for the M2M market, stating that it believes M2M will become an important mass market. This is just the latest announcement of a series of recent initiatives in the M2M market, both in the United States and in Europe.

In April 2011, Ericsson announced the acquisition of longtime M2M platform provider Telenor Connexion, while in July TeliaSonera announced that it had signed a cooperation agreement with France Telecom-Orange and Deutsche Telekom to increase the quality of service and interoperability for M2M services. In May 2011, T-Mobile USA announced that it had cast off its M2M operational business to longtime service partner Raco Wireless, although in July T-Mobile USA struck a partnership with asset protection provider IContain and Asset Protection Products LLC to help reduce operating costs of $7 billion in the US rent-to-own (RTO) sector.

Those and other initiatives signal that the M2M market is deemed ready to truly become a mass market, and players from hardware providers to M2M specialists passing through telco operators and sytem integrators [208] are trying to position themselves to reap the benefits.

While the executive-level comments and business unit launches from AT&T and Verizon signal a highly promising vision for the future, the reality of the M2M market is different and less optimistic as seen by other analysts such as Berg Insights. A comparison of analyst projections for the M2M market points to a market of about 100 million unit shipments for 2012 [38]. Strategy Analytics identifies five key barriers to scaling the global M2M market [275]:

1. Lack of a low-cost local access media that can be implemented on a global basis
2. The fragmented nature of both the technology vendors and the solutions they provide

3. Lack of any single killer application that can consolidate the market and drive demand forward
4. The increased costs associated with development and integration because of the complex nature of M2M solutions
5. Management's inability to express the benefits of M2M in anything other than cost savings, rather than exploiting and encouraging the service enablement capacity of mobile M2M

Figure 3.3 shows the typical architecture of an M2M system from BiTX. The integration middleware at the server side is the brain of the entire system.

Cellular networks were designed for circuit-switched voice. While they do a perfectly adequate job for regular, packet-switched data such as email and web browsing, they do not have the requisite functionality for M2M applications. For example, the normal OSS (operation support system) and BSS (business support system) are not designed for low-cost, mass handling of huge amounts of similar subscriptions. That led to the development of service enablement middleware platforms by specialized service providers (Table 3.3).

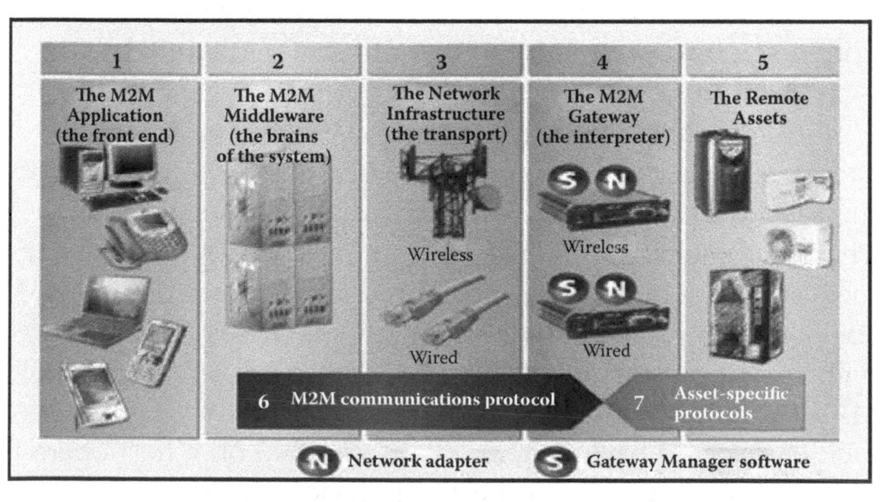

Figure 3.3 BiTX M2M architecture based on middleware.

Table 3.3 M2M Service Enablement Middleware

Vertical Applications Applications to connect to and communicate with objects tailored for specific verticals. Must be done in partnership with industry.
Service Enablement Middleware (APIs over Internet) Reduce complexities with regard to fragmented connectivity, device standards, application information protocols, etc., and device management. Build on and extend connectivity.
Connectivity (ADSL, SMS, USSD, GSM, GPRS, UMTS, HSPA, WiFi, Satellite, Zigbee, RFID, Bluetooth, etc.) Connectivity tailored for object communication with regards to business model, service level, SIM provisioning, billing, etc.

Service enablement is a middleware layer that facilitates the creation of applications. You can think of it as an operating system that the software developers write to this layer via application programming interface (APIs). A significant percentage of the functionality of the middleware comes from the charging, mediation, service management, and network management solutions that are being deployed in next-generation networks. These components have functionality that is similar and in some ways superior to that of regular M2M middleware platforms.

Table 3.4 shows the value chain of M2M business, which can be separated into two parts: the first relating to devices and the second to application development and service delivery. The broad intersection between these two parts represents the means by which devices are procured and integrated into M2M solutions and services. Both MNOs (mobile network operators), with some operators taking a more active role than others, and MVNOs (mobile virtual network operators, as shown in the table), subject to having their devices certified on a host operator's network, are trying to be M2M service providers.

The M2M device market share of chipset vendors including TI, Infineon, ST-Ericsson, Qualcomm, and others, and module

Table 3.4 Operator's Participation in Value Chain

Activity in Value Chain / Type of Entity	*Mobile Network Operator*	*Mobile Network Enabler*	*Mobile Virtual Network Enabler*	*Mobile Virtual Network Operator*	*Branded Reseller*	*Service Provider*
Mobile License	X	X				
Mobile Infrastructure	X	X				
Direct Customer Relationship	X			X	X	X
Network Routing	X	X	X	X		
Roaming Agreements	X	X	X	X		
Customer Services Delivery	X	X	X	X		X
Billing	X	X	X	X		X
Mobile Handset Management	X	X	X	X		X

vendors including Enfora, Infone, Kyocera, Murata, Mobicom, Novatel, Panasonic, Semco, Siemens, Sierra Cellular, Simcom, Telit, Wavecom, and others, can be found in [209].

As MNOs become more directly involved with M2M application service providers (ASPs), some MNOs such as Sprint, AT&T, Verizon Wireless, China Mobile, China Telecom, China Unicom, Orange, Rogers Communications, Telenor, Telefonica, NTT DOCOMO, and others are actively deploying M2M-based services. Many are deploying key network elements, specifically mobile packet gateways (e.g., Gateway GPRS Support Node [GGSN], Packet Data Serving Node [PDSN], Home Location Register [HLR], etc.), specifically for their M2M operations, separate from their general mobile data infrastructure. Key benefits of doing this are that it simplifies internal business operations and optimizes use of the network.

Likewise, MVNOs active in the M2M market are also increasingly deploying mobile packet gateways and similar equipment to interconnect with their MNO partners' radio infrastructure. (ABI Research classifies MVNOs who have deployed HLRs and mobile packet gateways as "MMOs" [52]; i.e., M2M Mobile Operators, Aeris Communications, Jasper Wireless, Numerex, Kore Telematics, Wyless, Qualcomm nPhase, Wireless Maingate, etc., are examples of MMOs.) The benefits to the MVNO for doing this include the ability to create new service offerings independently of their MNO partners and to enable quicker provisioning and diagnostic capabilities to their ASP customers.

MMOs and ASPs are called M2M partners of MNOs. They could use only the connectivity services of an MNO or other services such as rating and charging. Amazon eReaders, M2M DataSmart, FleetMatics, TeloGis, and others are examples of ASPs. Jasper Wireless is an example that uses less services of MNOs in some applications, because it's also an MMO.

As more and more MNOs start to enter into the M2M market directly, such as Telenor Objects, etc., some ASPs and MMOs are forced to become mobile virtual network enablers

(MVNEs), that is, MNO or MVNO enablers for M2M. For example, Jasper Wireless is an MVNE of some of AT&T's M2M businesses.

There is virtually no MVNO in existence in China because there is no regulation allowing such a business or service; the Big Three state-owned telcos, China Mobile, China Unicom, and China Telecom, dominate the market. Based on the flagship product ^{ez}M2M Middleware Platform for IoT applications, built at THTF Co., Ltd. (the second largest system integrator of China) led by the author, THTF has successful established a joint venture with China Mobile to construct the M2M Platform for China Mobile's M2M/IoT base in ChongQing serving nationwide users for all vertical applications.

3.3 RFID: The Internet of Objects

The term *Internet of Things* was first used by Kevin Ashton, co-founder and executive director of the Auto-ID Center, when he was doing RFID-related research at Massachusetts Institute of Technology in 1999. The Auto-ID lab is a research federation in the field of networked RFID and emerging sensing technologies, consisting of seven research universities located on four different continents chosen by the former Auto-ID Center to design the architecture for the Internet of Things together with EPCglobal. The technology they have developed is at the heart of a proposal sponsored by EPCglobal and supported by GS1, GS1 US, Walmart, Hewlett-Packard, and others to use RFID and the electronic product code (EPC) in the identification of items in the supply chain for companies.

An RFID tag is a simplified, low-cost, disposable contactless smartcard. RFID tags include a chip that stores a static number (ID) and attributes of the tagged object and an antenna that enables the chip to transmit the store number to a reader. When the tag comes within the range of the appropriate RF reader, the tag is powered by the reader's RF

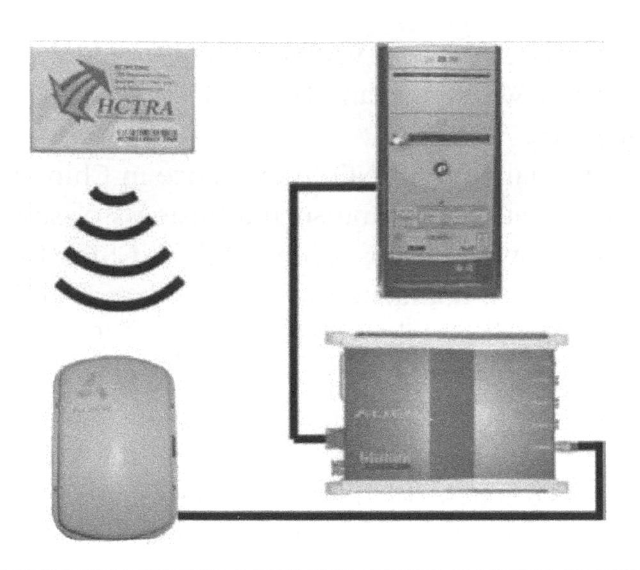

Figure 3.4 RFID system components. (From Erick C. Jones and Christopher A. Chung, *RFID in Logistics: A Practical Introduction,* Boca Raton, FL: CRC Press, 2008.)

field and transmits its ID and attributes to the reader. The contactless smartcard provides similar capabilities but stores more data.

An RFID system involves hardware known as readers and tags, as well as RFID software or RFID middleware (Figure 3.4). RFID tags can be active, passive, or semipassive. Passive RFID does not use a battery, while an active has an on-board battery that always broadcasts its signal. A semipassive RFID has a small battery on board that is activated when in the presence of a RFID reader.

The RFID technology is different from the other three technologies of IoT in the sense that it tags on an "unintelligent" object such as a pallet or an animal (an early experiment with RFID implants was conducted by British professor of cybernetics Kevin Warwick, who implanted a chip in his arm in 1998) to make it an instrumented [180] intelligent object for monitoring and tracking, while the other three (M2M, WSN, and Smart Systems) simply connect "intelligent" electronic devices.

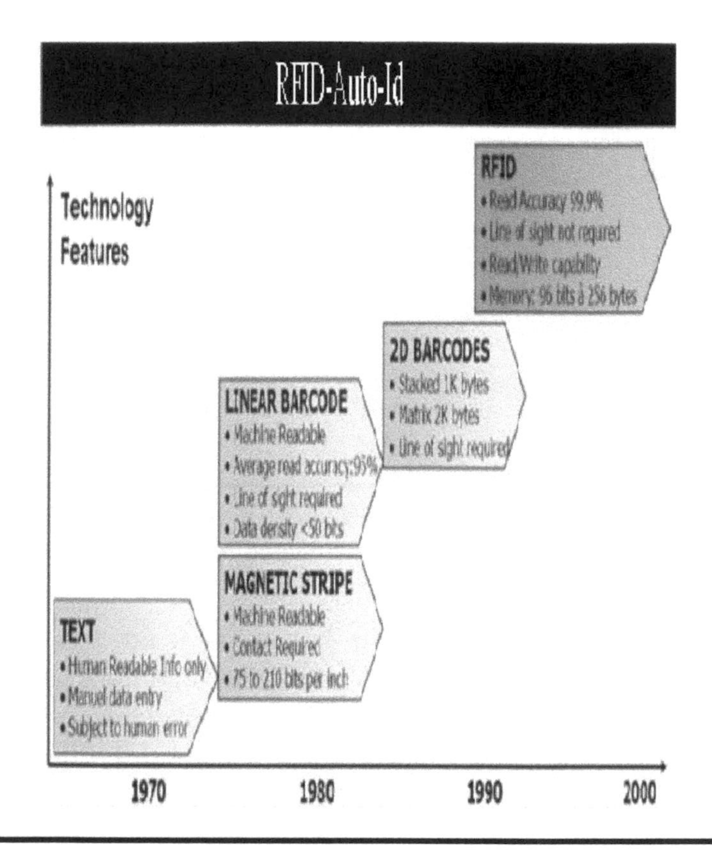

Figure 3.5 Evolution of identifications.

Mario Cardullo's passive radio transponder device in 1973 was the first true ancestor of modern RFID. For object or article identifications, text and then barcodes were widely used before RFID tags come into being (Figure 3.5).

UPC (universal product code) of UCC (Uniform Code Council, later called GS1 US) was widely used in the United States and Canada for tracking trade items in stores (Figure 3.6). EAN (European article number), developed after UPC, was used in Europe. EAN International is now called GS1. All the numbers encoded in UPC and EAN (as well as EAN/UCC-13, EAN/UCC-14, EAN-8, etc.) bar codes are known as global trade item numbers (GTIN). GS1, GS1 US, and Auto-ID labs joined forces to form EPCglobal in 2003 (which means the United States and Europe share the EPC standard; however, UID

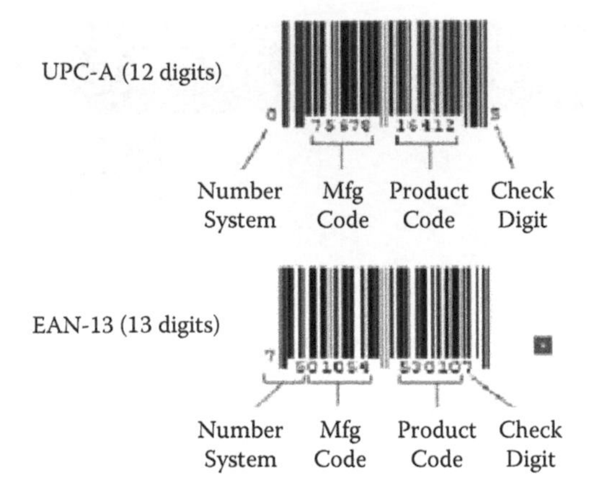

Figure 3.6 Bar code formats. (From James B. Ayers and Mary Ann Odegaard, *Retail Supply Chain Management*, New York: Auerbach Publications, 2008.)

[ubiquitous ID] is used in Japan). EPCglobal is an organization set up to achieve worldwide adoption and standardization of EPC technology. The main focus of the group currently is to create both a worldwide standard for RFID and the use of the Internet to share data via the EPCglobal Network™.

The automotive industry has been using the technology in manufacturing for decades. Pharmaceutical companies are already adopting the technology to combat counterfeiting. The Department of Homeland Security has been looking to leverage RFID along with other sensor networks to secure supply chains and ensure port and border security. Many major businesses already use RFID for better asset visibility and management. But the RFID technology and applications became widely used after the industry mandates started in 2004. Walmart and the U.S. Department of Defense (DOD) along with some other major retailers required their suppliers to begin RFID tagging pallets and cases shipped into their distribution centers in 2005 (http://www.controlelectric.com/RFID/Wal-Mart_DOD_Mandates.html). The mandates impacted some 200,000 suppliers globally. That year was also when the ITU published the Internet of Things report. Many companies

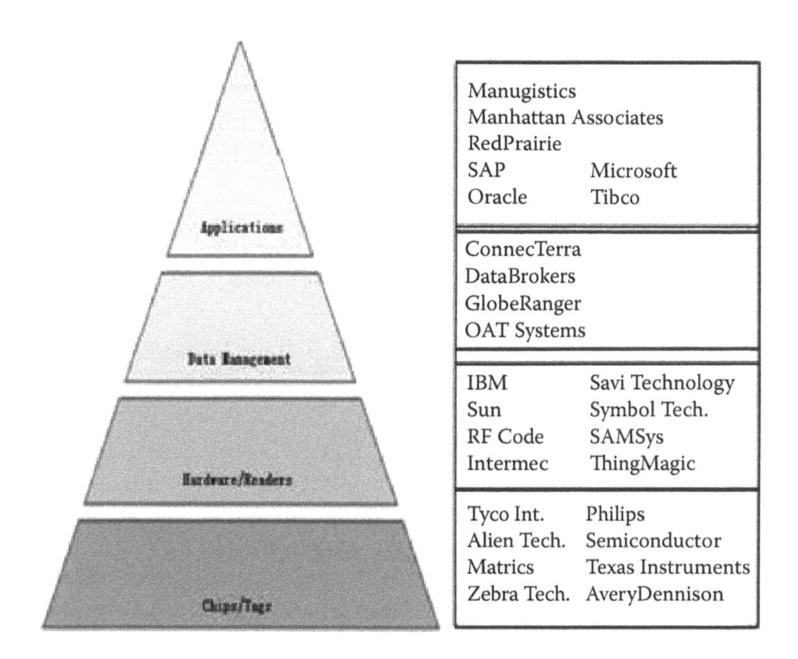

Figure 3.7 RFID value chain and vendors.

worldwide have since started to aggressively invest and build RFID technologies and products. Figure 3.7 shows a list of RFID vendors and solutions introduced in 2004.

The International Organization for Standardization asserts jurisdiction over the air interface for RFID through standards-in-development ISO 18000-1 through 18000-7. These are represented in the United States by American National Standards Institute and the Federal Communications Commission. The frequencies available are shown in Table 3.5.

The Auto-ID concept is that the data will be stored on the Internet or the EPCglobal network, and the EPC stored in the tag is used as an index to locate the data. This introduces several standards as shown in the EPCglobal architecture framework [51], which is a collection of interrelated standards for hardware, software, and data interfaces, together with core services that are operated by EPCglobal and its delegates.

All the software specifications from the Auto-ID Center are written in and for Java. Java-based middleware plays

Table 3.5 RFID Frequency Ranges

RFID	*Key Applications*	*Standard*
125 kHz (LF)	Inexpensive passive RFID tags for identifying animals	ISO 18000-2
13.56 MHz (HF)	Inexpensive passive RFID tags for identifying objects; library book identification, clothes identification, etc.	ISO 14443
400 MHz (UHF)	For remote control for vehicle center locking systems	ISO 18000-7
868 MHz, 915 MHz, and 922 MHz (UHF)	For active and passive RFID for logistics in Europe, the United States, and Australia, respectively	Auto-ID Class 0 Auto-ID Class 1 ISO 18000-6
2.45 GHz (MW)	An ISM band used for active and passive RFID tags; e.g., with temperature sensors or GPS localization	ISO 18000-4
5.8 GHz (MW)	Used for long-reading range passive and active RFID tags for vehicle identification, highway toll collection	ISO 18000-5

an important and pivotal role in the implementation of the EPCglobal architecture framework, especially the application level events (ALE) and EPC information services (EPCIS). That's why middleware and software giants such as IBM, Oracle, Microsoft, and SAP all have large investments in RFID and developed complete RFID solution stacks.

The ONS (object naming service) is an authoritative directory service just like the DNS (domain name service) for the Internet that routes requests for information about EPCs between a requesting party and the product manufacturer, via a variety of existing or new network- or Internet-based information resources. That's why EPCglobal has worked with VeriSign to provide such a service in addition to VeriSign's

DNS. VeriSign has operated the authoritative root directory for the EPCglobal Network since 2005. Although companies have successfully implemented internal RFID solutions that have captured efficiencies within the enterprise, the greatest promise of the EPCglobal Network is the ability to extend the benefits across trading-partner boundaries via the Internet to realize the IoT vision. It is not hard to imagine that RFID can be used in almost all industry segments and the benefits it will bring.

There are many estimates of the RFID market size. IDTechEx predicts that the total market of RFID will be around US$27 billion worldwide in 2018. The market size of China will be around US$1.7 billion in 2014 per iSuppli reports [210]. The RFID market was more than US$3 billion in 2008 in China when the issuing of RFID-based national ID cards for each citizen reached its peak.

In a contactless smart card, using NFC (near field communication) technologies, the chip communicates with the card reader through an induction technology similar to that of RFID. These cards require close proximity to an antenna to complete a transaction. They are often used when transactions must be processed quickly or hands-free, such as on mass transit systems, where a smart card (ticket) can be used without even removing it from a wallet. Figure 3.8 shows the RFID-based ticket and the ^{ez}M2M middleware-based application system the author's team built for the Beijing Olympic Games in 2008.

Mobile payment or mobile wallet is an alternative payment method that has been well adopted in many parts of Europe and Asia. Juniper Research forecasts that the combined market for all types of mobile payments is expected to reach more than $600 billion globally by 2013. RFID/NFC technologies have been used for mobile payments in China by its big three telco companies as well as China UnionPay, whose UnionPay cards can be used in 104 countries and regions around the world.

Figure 3.8 Example of RFID application.

3.4 WSN: The Internet of Transducers

As defined in the first section, WSN is more for sensing and information-collecting purposes. Other networks include BSN (body sensor network [56]), VSN (visual or video sensor network [54,55]), vehicular sensor networks (V2V, V2I), underwater (acoustic) sensor networks (UW-ASN), urban/social/participatory sensor networks, interplanetary sensor networks, fieldbus networks (categorized as SCADA systems, the good oldies in the buildings and plants are getting wireless/mobile capabilities and scaling up), and others.

BSN is a term used to describe the application of wearable computing devices to enable wireless communication between several miniaturized body-sensor units and a single body central unit worn on the human body to transmit vital signs and motion readings to medical practitioners or caregivers (Figure 3.9). Applications of BSN are expected to appear primarily in the healthcare domain, especially for continuous monitoring and logging of vital parameters for patients suffering from chronic maladies such as diabetes, asthma, and heart attacks.

Visual sensor networks are based on several diverse research fields, including image/vision processing, communication and networking, and distributed and embedded system

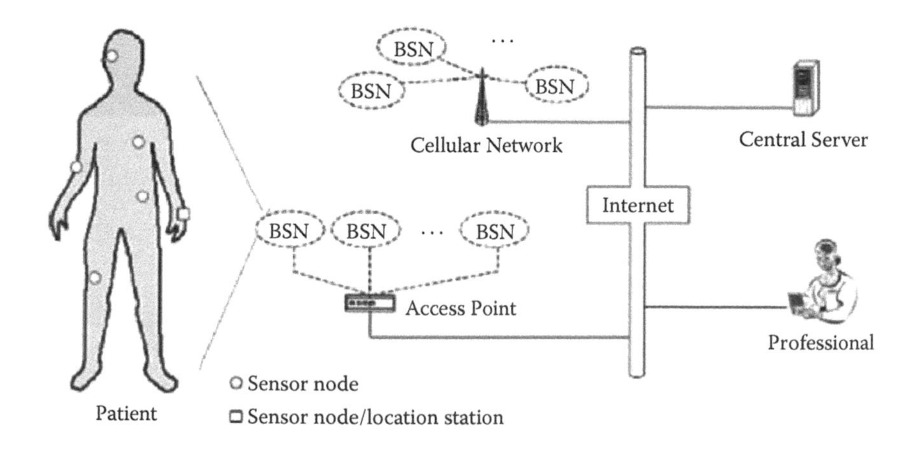

Figure 3.9 Body sensor networks. (From Hui Chen and Yang Xiao (eds.), *Mobile Telemedicine: A Computing and Networking Perspective*, New York: Auerbach Publications, 2008.)

processing. Applications include surveillance, environmental monitoring, smart homes, virtual reality, and others.

With the development of WSN, recent technological advances have led to the emergence of distributed wireless sensor and actuator networks (WSANs) that are capable of observing the physical world, processing the data, making decisions based on the observations, and performing appropriate actions. These networks can be an integral part of systems such as battlefield surveillance and microclimate control in buildings; nuclear, biological and chemical attack detection; home automation; and environmental monitoring.

The extended scope of WSN is the USN, or ubiquitous sensor network, a network of intelligent sensors that could one day become ubiquitous [53]. This USN is also a unified "invisible," "pervasive," or "ambient intelligent" Internet of Things.

The development of WSNs was motivated by military applications such as battlefield surveillance. The WSN is built of nodes—from a few to several hundred or even thousands—each node connected to one (or sometimes several) sensors. Each such sensor network node has typically several parts: a radio

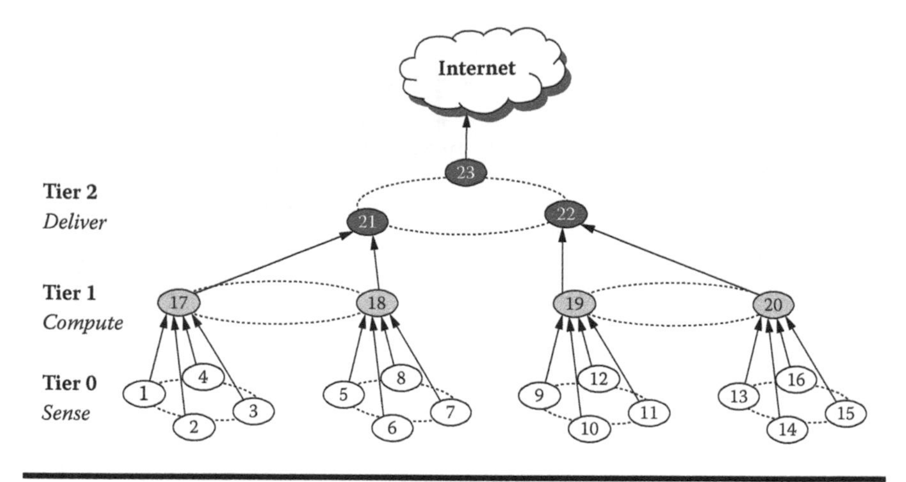

Figure 3.10 Sensor network architecture. (From Mark Yarvis and Wei Ye, "Tiered Architectures in Sensor Networks," in Mohammad Ilyas and Imad Mahgoub (eds.), *Handbook of Sensor Networks: Compact Wireless and Wired Sensing Systems*, Boca Raton, FL: CRC Press, 2004.)

transceiver with an antenna, a microcontroller, an electronic circuit for interfacing with the sensors, and an energy source, usually a battery or an embedded form of energy harvesting.

The architecture of a typical sensor network is shown in Figure 3.10. The topology of the WSNs can vary from a simple star network to an advanced multihop mesh network with a gateway sensor (sink) node connected (e.g., via a cellular M2M module) with a remote central server.

- Sensor node: sense target events, gather sensor readings, manipulate information, send them to gateway via radio link
- Base station/sink: communicate with sensor nodes and user/operator
- Operator/user: task manager, send query

Routing is required for reliable data transmission in a WSN mesh network. Routing protocols are distributed and reactive: nodes in the system start looking for a route only when they have application data to transmit. Ad hoc on-demand distance vector (AODV) and dynamic source routing (DSR) are frequently used routing algorithms.

The U.S. DOD, which operates the largest and most complex supply chain in the world, awarded in January 2009 a contract for $429 million in DASH7 infrastructure. This represents a major development in terms of global adoption of an ultra-low-power WSN technology based on a single global standard [72].

WSN is currently an active research area with limited mission-critical uses. IT giants such as IBM and Microsoft have invested in WSN research for a long time with little commercial success. Currently there is no common WSN platform. Some designs such as Berkeley Motes and their clones have broader user and developer communities. However, many research labs and commercial companies prefer to develop and produce their own devices. Since there is no true killer application for WSNs that would drive the costs down, it is often more convenient and even less expensive to build your own WSN devices than to buy commercially available ones.

Some of the existing WSN platforms are summarized in Table 3.6. Most of the device designs are still in the research stage.

According to IDTechEx, the price per WSN node was about $30 in 2011. In the future (10 years), a functionally equivalent "smart dust" sensor node is expected to be available for use with cost per node less than $1.

Energy is the scarcest resource of WSN nodes, and it determines the lifetime of WSNs. WSNs are meant to be deployed in large numbers in various environments, including remote and hostile regions, with ad hoc communications as key. For this reason, algorithms and protocols need to address the following issues:

- Lifetime maximization
- Robustness and fault tolerance
- Self-configuration

Table 3.6 RFID Platforms

Accsense, Inc. (http://www.accsense.com/)
Ambient Systems mesh networks (Netherlands) (http://www.ambient-systems.net/ambient/technology-features.htm)
Atlas (Pervasa/University of Florida) (http://www.pervasa.com/)
BEAN Project (http://www.dcc.ufmg.br/~mmvieira/publications/bean.pdf#search=%22BEAN%20brazilian%20sensor%20node%22)
Berkeley Motes/Piconodes
BTnode (ETH Zurich) (http://www.btnode.ethz.ch)
Cortex Project
COTS Dust (Dust Networks) (http://www.dustnetworks.com/
EYES Project (http://www.eyes.eu.org)
Fleck (CSIRO Australia) (http://www.btnode.ethz.ch/Projects/Fleck)
Glacsweb from University of Southampton (http://www.glacsweb.org)
G-Node from SOWNet Technologies (http://sownet.nl/index.php/en/products/gnode)
Global Sensor Networks (http://gsn.sourceforge.net/)
Hoarder Board—Open Hardware Design (MIT Media Lab) (http://vadim.oversigma.com/Hoarder/Hoarder.htm)
iSense hardware platform from Coalesenses GmbH, Germany (http://www.coalesenses.com)
Kmote (TinyOS Mall) (http://www.tinyosmall.co.kr/)
MeshScape (Millennial Net, Inc.) (http://millennialnet.com/Technology.aspx)
Mica Mote (Crossbow) (http://www.xbow.com/Products/productsdetails.aspx?sid=62)
MicroStrain, Inc. (http://www.microstrain.com/)
Newtrax Technologies, Inc. (http://www.newtraxtech.com/)
openPICUS—Open Hardware (http://openpicus.blogspot.com/)

Table 3.6 (continued) RFID Platforms

Particles (Particle Computer) spun out of TecO, Univ. of Karlsruhe) (http://www.particle-computer.de
PicoCrickets (Montreal, Canada) (http://www.picocricket.com)
Redwire Econotag (http://www.redwirellc.com/store/node/1)
ScatterWeb ESB nodes (http://www.inf.fu-berlin.de/inst/ag-tech/scatterweb_net/)
SensiNet Smart Sensors (Sensicast Systems) (http://www.sensicast.com)
Sensor Internet Project (http://sip.deri.ie)
Sensor Webs (SensorWare Systems) spun out of the NASA/JPL Sensor Webs Project (http://www.sensorwaresystems.com/)
Shockfish TinyNodes
Smart Dust (Dust Networks) spun out of UC Berkeley (http://www.dustnetworks.com/)
TIP Mote (Maxfor) (http://www.maxfor.co.kr/)
Tmote (Moteiv) spun out of UC Berkeley (http://www.moteiv.com/)
Tyndall Motes (http://www.tyndall.ie/mai/Wireless%20Sensor%20Networks.htm)
UCLA iBadge
Waspmote (Libelium) (http://www.libelium.com/waspmote)
WINS (Rockwell Wireless Integrated Network Sensors)
WINS (UCLA)
WSN430 (INSA de Lyon/INRIA) (http://www.senslab.info/)
XYZ node (http://www.eng.yale.edu/enalab/XYZ/)

WSNs have found more and more applications in a variety of pervasive computing environments. However, how to support the development, maintenance, deployment and execution of applications over WSNs remains a nontrivial and challenging task, mainly because of the gap between the

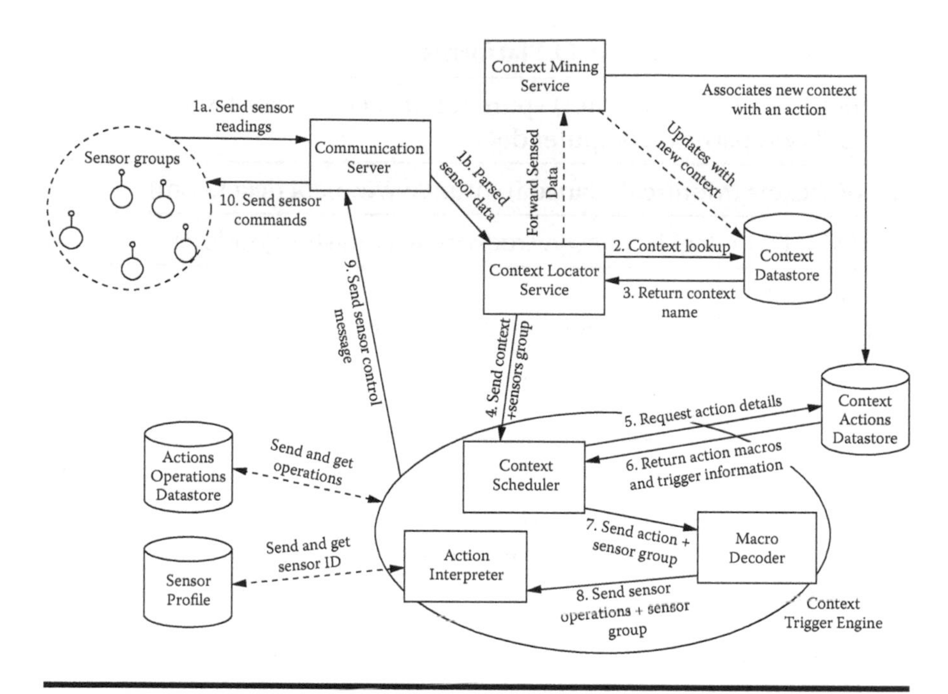

Figure 3.11 Context-aware system based on WSN. (From Seng Loke, *Context-Aware Pervasive Systems: Architectures for a New Breed of Applications,* New York: Auerbach Publications, 2007.)

high-level requirements from pervasive computing applications and the underlying operation of WSNs. Middleware for WSN, the middle-level primitives between the software and the hardware, can help bridge the gap and remove impediments. Middleware can help build context-aware IoT systems as shown in Figure 3.11.

Mobile sensor networks (MSNs) are WSNs in which nodes can move under their own control or under the control of the environment. Mobile networked systems combine the most advanced concepts in perception, communication, and control to create computational systems capable of interacting in meaningful ways with the physical environment, thus extending the individual capabilities of each network component and network user to encompass a much wider area and range of data. A key difference between a mobile WSN and a static WSN is how information is distributed over the network. Under static nodes, a new task or data can be flooded across the network in a very predictable way. Under mobility this kind of flooding is more complex, depending on the mobility model of the nodes in the system. The proliferation of commodity smartphones that can provide location estimates using a variety of sensors—GPS, WiFi real-time locating systems (RTLS), or cellular triangulation—opens up the attractive possibility of using position samples from drivers' phones to monitor traffic delays at a fine spatiotemporal granularity. MSN systems such as vTrack [58] of the MIT CarTel group have been built to monitor traffic delays and change routes.

According to IDTechEx research in the new report "Wireless Sensor Networks 2011–2021" [211], the WSN market is expected to grow rapidly from $0.45 billion in 2011 to $2 billion in 2021. These figures refer to WSN defined as wireless mesh networks, that is, self-healing and self-organizing. WSNs will eventually enable the automatic monitoring of forest fires, avalanches, hurricanes, failure of country-wide utility equipment, traffic, hospitals, and much more over wide areas, something previously impossible. More humble killer applications already exist such as automating meter readings in buildings, and manufacture and process control automation.

The United States dominates (72 percent, according to IDTechEx, of all countries worldwide) the development and use of WSN partly because of the heavier funding available. The U.S. WSN industry sits astride the computer industry thanks to companies such as Microsoft and IBM, and WSN is regarded as a next wave of computing, so U.S. industry is particularly interested in participating. Add to that the fact that the U.S. military, deeply interested in WSN, spends more than all other military forces combined, and creating and funding start-ups is particularly easy in the United States, and you can see why the United States is ahead at present.

3.5 SCADA: The Internet of Controllers

For more than a decade, many in the building industry have been envisioning a day when building automation systems (BAS) would become fully integrated with communication and human interface practices and standards widely employed for information technology systems. Not long ago, building automation graphical interfaces (shown in Figure 3.12; the part on the right is the human–machine interface the author's team built for the super-energy-efficiency building at QingHua University) employed almost no web-browser techniques and technologies; now, web approaches are the basis of many such packages. How close we are to a complete convergence of BAS and IT is difficult to tell, but it is not too much of a stretch to say that when the convergence is complete, there may be nothing to distinguish one from the other [59].

SCADA (supervisory, control and data acquisition) systems, as the core technology of the controls–IT convergence, will evolve and take the center stage. By their very nature, SCADA, low-data-rate (LDR), and M2M/IoT [129] services are closely related and largely overlapped in technologies and deployment approaches, as per GII Research [60]. Also, WSN is considered a new computing paradigm that emerged from the fusion of the SCADA systems and ad hoc networks technologies [61]. The advent of the Internet of Things will no doubt speed up the controls–IT convergence and make control systems and IT systems inseparable and indistinguishable from each other.

SCADA was generally referring to industrial control systems (ICSs): computer systems that monitor and control industrial, infrastructure, or facility-based processes, as described below:

- Industrial processes include those of manufacturing, production, power generation, fabrication, and refining, and may run in continuous, batch, repetitive, or discrete modes.

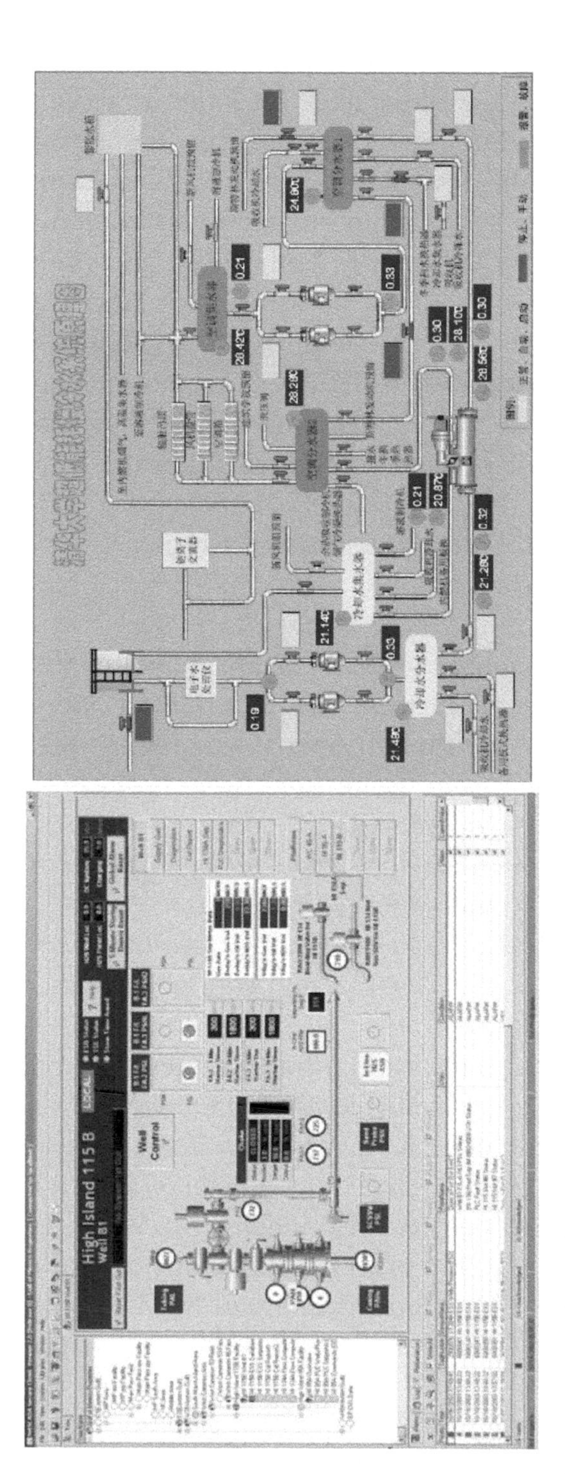

Figure 3.12 Examples of SCADA graphics and animations.

- Infrastructure processes may be public or private and include water treatment and distribution, wastewater collection and treatment, oil and gas pipelines, electrical power transmission and distribution, wind farms, civil defense siren systems, and large transportation systems.
- Facility processes occur in both public and private facilities, including buildings, airports, ships, and space stations. They monitor and control HVAC, access, and energy consumption using PLCs (programmable logic controllers) and DCSs (distributed control systems) via the OPC (OLE for process control) middleware.

An existing SCADA system usually consists of the following subsystems (Figure 3.13):

- A human–machine interface (HMI), which is the apparatus that presents process data to a human operator, and through this, the human operator monitors and controls the process.
- Remote terminal units (RTUs) connect to sensors in the process, convert sensor signals to digital data, and send digital data to the supervisory system.
- PLCs are used as field devices because they are more economical, versatile, flexible, and configurable than special-purpose RTUs.
- DCSs; as communication infrastructures with higher capacity become available, the difference between SCADA and DCS will fade. SCADA is combining the traditional DCS and SCADA.
- As mentioned before, M2M (telemetry), WSN, smart systems, CPS, and others all have overlaps of scope with SCADA, but the extended scope of SCADA is bigger under the IoT umbrella.

A SCADA system could be a layer between the top-layer business systems such as ERP, WMS (warehouse management

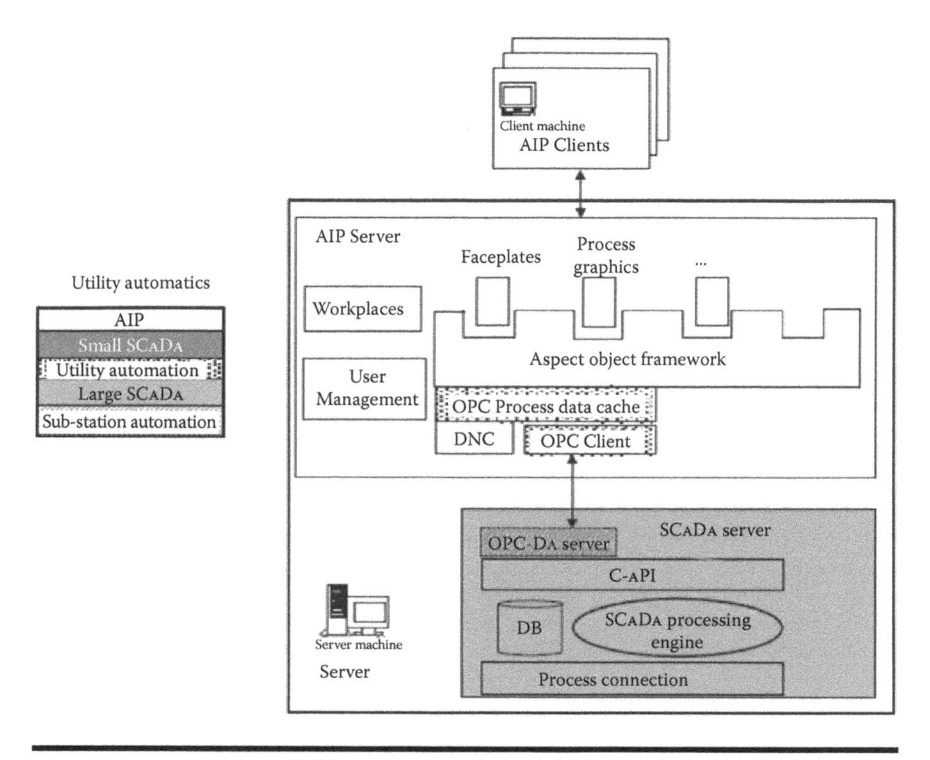

Figure 3.13 Components of a SCADA system. (From Yauheni Veryha and Peter Bort, "Industrial IT-Based Network Management," in Richard Zurawski (ed.), *The Industrial Information Technology Handbook*, Boca Raton, FL: CRC Press, 2005.)

system), SCM, CRM, EAM (enterprise asset management), PIMS (plant information management system), EMI (enterprise manufacturing intelligence), LIMS (laboratory information management system), and other applications and the lower layer DCS, PLC, RTU, MES (manufacturing execution system), SIS (supervisory information system in plant level), and other systems as exemplified in Figure 3.14.

A traditional SCADA system is a client/server system. New technological developments have turned C/S SCADA systems into middleware-backed, web-based, three-tiered open systems with SOA capabilities.

Figure 3.15 showcases a typical SCADA middleware or platform architecture. Examples of such platforms include

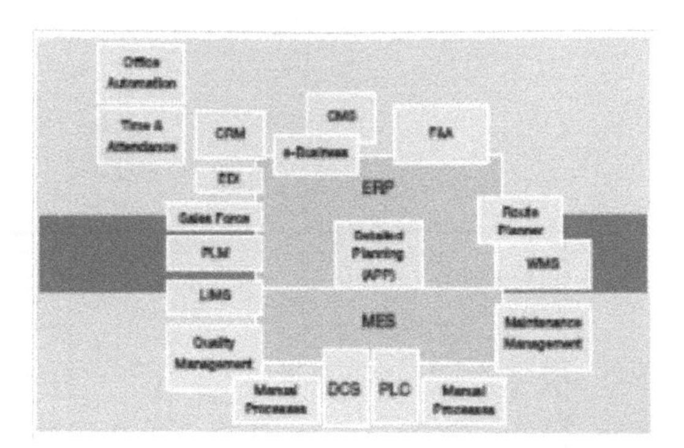

Figure 3.14 SCADA sits in the center.

Operating System (kerel, libraries, binaries etc.)
Linux, Nucleus, Symbian, WinCE, VxWorks, QNX

Application Layer

Information Systems
Visualization Applications
SCADA Packages
Remote Monitoring & Diagnostics
Systems Configurator
Safety Applications
Modeling & Simulations
Web & Server based Applications

Service and Application Manager

Device Management
Security Management
JAVA Based Applications

Middleware

Human Machine Interface (Touch, GUI, Tactile, Command Line)
Mobility, Connectivity, Profile, Content Management

Control Methods (Motion, Discrete, Process and Batch and Drive Controls)

Network Connectivity and Protocols (BACNet, Lonworks, Metasys, CAN OPC, HART, MODBUS, CIP)

Protocol Stack

Test & Measurements Solutions
Motion & Drives Control Solutions
Industrial & Control Solutions

BSP, Device Drivers (Power Mgmt., Keyboard, USB, Camera, LCD, UART, SPI, i2c, etc.)

Hardware

Figure 3.15 Middleware-based SCADA systems.

(Invensys) Wonderware's ArchestrA™, (Honeywell) Tridium's Niagara Framework™ (a Java EE–based platform), THTF's ezM2M Middleware for IoT, various implementations of the OPC UA framework standard, and the list goes on.

SCADA systems allow the automation of complex industrial processes where human control is impractical. However, with all the raw data and real-time updates pouring in, it can be difficult to decipher what is going on and how to respond. All the on-screen numbers, flashing lights, and blaring alarms still leave you in the dark. The solution is an integrated controls–IT convergence system [59,183].

IP video technology has become one of the hottest trends in the automation industry today, especially since automation and surveillance systems have both migrated to IP-based applications. Moreover, the integration of IP surveillance software with automation systems is gaining popularity and momentum, and integrating real-time visual surveillance systems [100] with SCADA systems via IP video technology is now both a viable and an affordable solution for system integrators.

Many industries are using SCADA as a core technology to link the geographically separated facilities and support new business processes in response to changing industry dynamics.

As examples, the worldwide oil and gas industry SCADA market was $850 million in 2007 and is forecast to be over $1.3 billion in 2012; the worldwide market for electric power SCADA was $1.629 trillion in 2008 and is forecast to be over $2.125 trillion in 2013; and the worldwide water and wastewater industry SCADA market was $212 million in 2006 and is forecast to be over $275 million in 2011, all according to ARC Advisory Group studies.

In 2010, Chinese government and industry leaders stated that a "unified strong and smart grid" [166] system is going to be built across the country by 2020. SCADA sales will increase as part of this initiative and overall IoT development.

Supported by intelligent field devices, expanded communications networks, and improved compatibility with IT, especially the Internet and web technologies, SCADA can now provide a wealth of information and knowledge as a means to modify business processes and enable the creation of new SCADA-based IoT applications.

3.6 Summary

Many IoT technologies and applications are not new. IoT is an aggregation, convergence, and evolution of existing ICT technologies and applications. What should be included in the IoT paradigm has long been and still is an issue of many disagreements.

In this chapter, a solution to this disagreement is introduced. The four-pillar classification of the Internet of Things was proposed based on analysis of common IoT characteristics and previous categorization efforts. The technologies, applications, and market potentials of each pillar were described in detail.

In the next chapter, we will talk about the three DCM layers of IoT value chain, the role of each, and what is included in each layer.

Chapter 4

The DNA of IoT

4.1 DCM: Device, Connect, and Manage

The first issue that the Internet of Things (IoT) ecosystem needs to address is the long and fragmented value chain that characterizes the industry. This results in numerous supplier–buyer interfaces, adding costs and time to the launch of any new product offering.

Just like the blind men and the elephant story and people's understanding of the four pillars or the six pillars mentioned before, the IoT is still different things to different people, even though introduced more than a decade ago. However, there is one thing most people agree with: IoT (or machine-to-machine, M2M; wireless sensor networks, WSN; supervisory control and data acquisition, SCADA; radio-frequency identification, RFID; etc.) systems all have three layers. Figure 4.1 is an example IoT application of an intelligent nuclear power plant IoT system [63] of Datang Telcom in China. More examples of the three-layer architecture of IoT can be found at European Telecommunications Standards Institute (ETSI)'s website [212].

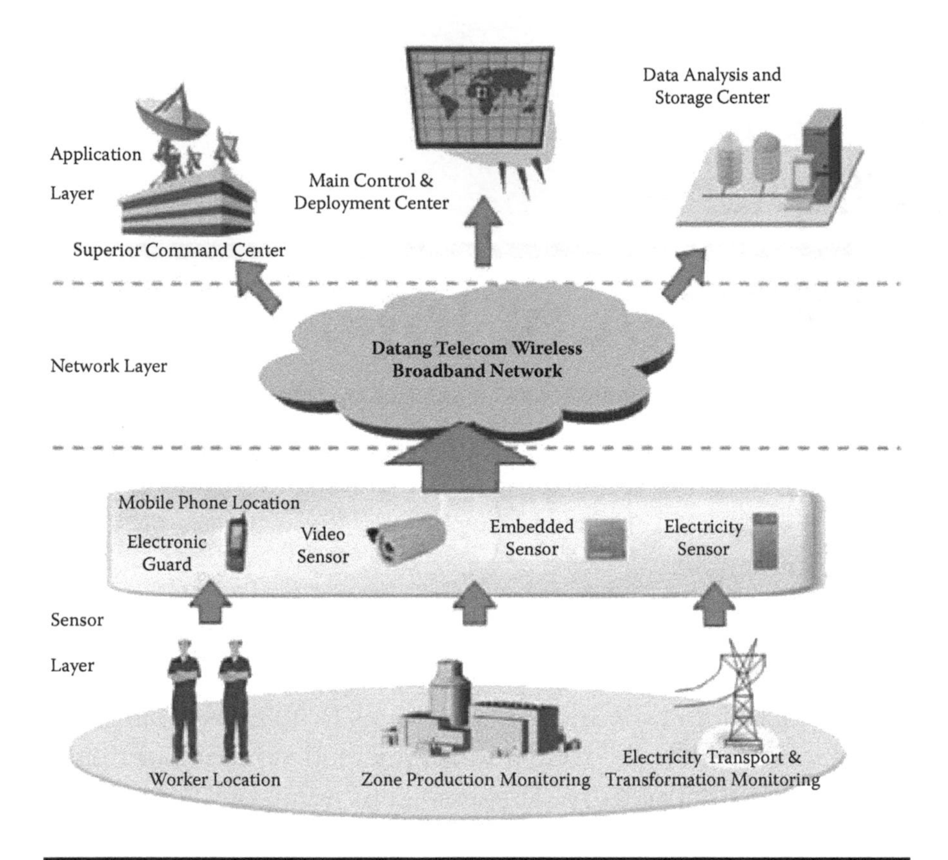

Figure 4.1 Examples of three-layer architecture of IoT.

The author has proposed the concept and acronym DCM (device, connect, and manage [74]) as a corporate stretegy or slogan for TongFang Co. Ltd. The board of the company announced financing of 500 million Chinese renminbi (RMB) (or US$78.5 million) for the development of the IoT/DCM business in 2005. Numerex created a better acronym called DNA™ (devices, networks, and applications) [213] in 2008 (Figure 4.2).

The three-layer DCM classification is more about the IoT value chain than its system architecture at runtime. For system architecture, some (e.g., one of Numerex's and IBM's reports) have divided the IoT system into as many as nine layers, from bottom to top: devices, connectivity, data collection,

M	• Vertical Applications • Server-side Middleware Platform • Data Management	A
C	• Machine Type Communication • Edge Middleware • Pervasive Networks	N
D	• Local/Ad-hoc Sensor Networks • Embedded Middleware • Sensors and Actuators	D

Figure 4.2 DCM (DNA) of IoT.

communication, device management, data rules, administration, applications, and integration.

While large companies such as IBM, Oracle, Microsoft, and others have comprehensive solutions, products, and services that cover almost the entire value chain, startups or smaller players in the IoT sector should focus on providing products or services in no more than two components or areas in the value chain. The following sections discuss the three DCM components.

4.2 Device: Things That Talk

According to the IoT definitions and descriptions in the previous chapters, devices or assets can be categorized as two groups: those that have *inherent intelligence* such as electric meters or heating, ventilation, and air-conditioning (HVAC) controllers, and those that are inert and must be *enabled* to become smart devices (e.g., RFID tagged) such as furniture or animals that can be electronically tracked and monitored—things that "talk."

Just as Paul Saffo [214], a technological forecaster and strategist, described in an interview in 2002:

> This is the Cambrian explosion of communications. We are seeing a radical species divergence

> of different kinds of devices and different types of things that want to talk, from your washing machine having an Internet connection and being able to scream for help if it is broken, to your car having a wireless connection for data telemetry back to the manufacturer. Today, voice communications is way below 1% of the total communications traffic on this planet. That's why people are giving voice away for free. So that means that we're going to see a whole zoo of new kinds of devices that have to talk. It's going to become a world of smartifacts, or intelligent objects. This stuff is so cheap, we're putting chips in everything, anything with a chip inside can be connected into the Internet of Things.

Devices that perform an input function are commonly called sensors because they "sense" a physical change in some characteristic that changes in response to some excitation, for example, heat or force, and convert that into an electrical signal. Devices that perform an output function are generally called actuators and are used to control some external device, for example, movement. Both sensors and actuators are collectively known as transducers because they are used to convert energy of one kind into energy of another kind. For example, a microphone (input device) converts sound waves into electrical signals for the amplifier to amplify, and a loudspeaker (output device) converts the electrical signals back into sound waves.

A *sensor* (also called a detector) is a device that responds to a physical stimulus, measures the physical stimulus quantity, and converts it into a signal, usually electrical, which can be read by an observer or by an instrument.

Based on this definition, a sensor is basically an electrical device. It could be an M2M terminal, an RFID reader, or a SCADA meter. Sensors are particularly useful for making in-situ measurements (things that talk) such as in industrial process control or medical applications. A sensor can be very small and itself can be a trackable device; however, when a train or an aircraft is instrumented with a small sensor, the entire aircraft becomes one trackable device.

The sensor itself, if not connected, is not part of the IoT or WSN value chain. This is like a central processing unit (CPU), which is not part of the web or social networking services, even though they are somwhat related. Some sensors do not generate electrical signals; for example, a mercury-in-glass thermometer converts the measured temperature into expansion and contraction of a liquid, which can be read on a calibrated glass tube. However, it's important to understand the types and shapes of the ubiquitous sensors if you are into IoT, just as an architect should know what concrete and cement are as well as their differences. Figure 4.3 showcases a few sample sensors.

Some of the existing sensors and their types are listed in Table 4.1. The size of the overall sensor market is difficult to estimate. A number of research reports on the market size of

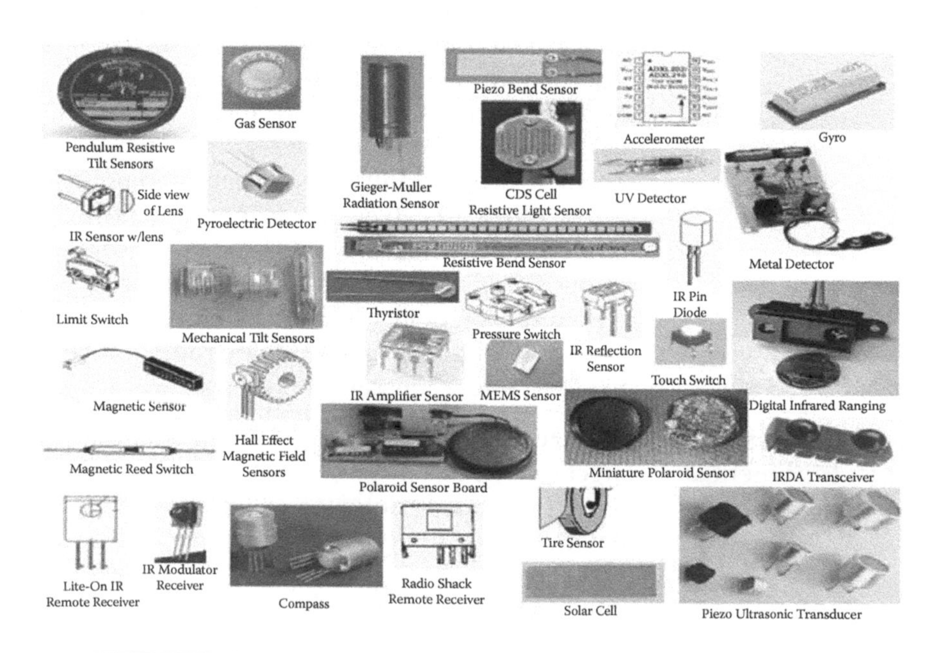

Figure 4.3 Examples of sensors.

Table 4.1 List of Sensors and Types

Sensor Type (Examples)	*Sensors (Examples)*
Acoustic, sound, vibration	Geophone, hydrophone, lace sensor, microphone, seismometer
Automotive, transportation	Air-fuel ratio meter, crank sensor, curb feeler, defect detector, engine coolant temperature (ECT) sensor, all effect sensor, MAP (manifold absolute pressure) sensor, mass flow sensor or mass airflow (MAF) sensor, oxygen sensor, parking sensors, radar gun, speedometer, speed sensor, throttle position sensor, tire-pressure monitoring sensor, transmission fluid temperature sensor, turbine speed sensor (TSS) or input speed sensor (ISS), ariable reluctance sensor, vehicle speed sensor (VSS), water sensor or water-in-fuel sensor, wheel speed sensor
Chemical	Breathalyzer, carbon dioxide sensor, carbon monoxide detector, catalytic bead sensor, chemical field-effect transistor, electrochemical gas sensor, electronic nose, electrolyte–insulator–semiconductor sensor, hydrocarbon dewpoint analyzer, hydrogen sensor, hydrogen sulfide sensor, infrared point sensor, ion-selective electrode, nondispersive infrared sensor, microwave chemistry sensor, nitrogen oxide sensor, olfactometer, optode, oxygen sensor, pellistor, pH glass electrode, potentiometric sensor, redox electrode, smoke detector, zinc oxide nanorod sensor
Electric current, electric potential, magnetic, radio	Ammeter, current sensor, galvanometer, hall effect sensor, hall probe, leaf electroscope, magnetic anomaly detector, magnetometer, metal detector, multimeter, ohmmeter, radio direction finder, telescope, voltmeter, voltage detector, watt-hour meter

Table 4.1 (continued) List of Sensors and Types

Sensor Type (Examples)	*Sensors (Examples)*
Environment, weather, moisture, humidity	Actinometer, bedwetting alarm, dew warning, fish counter, gas detector, hook gauge evaporimeter, hygrometer, leaf sensor, pyranometer, pyrgeometer, psychrometer, rain gauge, rain sensor, seismometers, snow gauge, soil moisture sensor, stream gauge, tide gauge
Flow, fluid velocity	Air flow meter, anemometer, flow sensor, gas meter, mass flow sensor, water meter
Force, density, level	Bhangmeter, hydrometer, force gauge, level sensor, load cell, magnetic level gauge, nuclear density gauge, piezoelectric sensor, strain gauge, torque sensor, viscometer
Ionizing radiation, subatomic particles	Bubble chamber, cloud chamber, geiger counter, neutron detection, particle detector, scintillation counter, scintillator, wire chamber
Navigation instruments	Air speed indicator, altimeter, attitude indicator, depth gauge, fluxgate compass, gyroscope, inertial reference unit, magnetic compass, MHD sensor, ring laser gyroscope, turn coordinator, variometer, vibrating structure gyroscope, yaw rate sensor
Optical, light, imaging, photon	Charge-coupled device, colorimeter, contact image sensor, electro-optical sensor, flame detector, infra-red sensor, kinetic inductance detector, LED as light sensor, Nichols radiometer, fiber-optic sensor, photodetector, photodiode, photomultiplier tubes, phototransistor, photoelectric sensor, photoionization detector, photomultiplier, photoresistor, photoswitch, phototube, scintillometer, Shack–Hartmann, single-photon avalanche diode, superconducting nanowire single-photon detector, transition edge sensor, visible light photon counter, wavefront sensor

continued

Table 4.1 (continued) List of Sensors and Types

Sensor Type (Examples)	*Sensors (Examples)*
Position, angle, displacement, distance, speed, acceleration	Accelerometer, auxanometer, capacitive displacement sensor, free fall sensor, gravimeter, inclinometer, laser rangefinder, linear encoder, linear variable differential transformer (LVDT), liquid capacitive inclinometers, odometer, piezoelectric accelerometer, position sensor, rotary encoder, rotary variable differential transformer, selsyn, sudden motion sensor, tilt sensor, tachometer, ultrasonic thickness gauge
Pressure	Barograph, barometer, boost gauge, bourdon gauge, hot filament ionization gauge, ionization gauge, McLeod gauge, oscillating U-tube, permanent downhole gauge, Pirani gauge, pressure sensor, pressure gauge, tactile sensor, time pressure gauge
Proximity, presence	Alarm sensor, Doppler radar, motion detector, occupancy sensor, proximity sensor, passive infrared sensor, reed switch, stud finder, triangulation sensor, touch switch, wired glove
Sensor technology	Active pixel sensor, biochip, biosensor, capacitance probe, catadioptric sensor, carbon paste electrode, displacement receiver, electromechanical film, electro-optical sensor, Fabry–Pérot interferometer, image sensor, inductive sensor, intelligent sensor, lab-on-a-chip, leaf sensor, machine vision, micro-sensor arrays, photoelasticity, RADAR, ground-penetrating radar, synthetic aperture radar, sensor array, sensor grid, sensor node, soft sensor, SONAR, underwater acoustic positioning system, staring array, transducer, ultrasonic sensor, video sensor, visual sensor network, Wheatstone bridge

Table 4.1 (continued) List of Sensors and Types

Sensor Type (Examples)	*Sensors (Examples)*
Thermal, heat, temperature	Bolometer, bimetallic strip, calorimeter, exhaust gas temperature gauge, gardon gauge, golay cell, heat flux sensor, infrared thermometer, microbolometer, microwave radiometer, net radiometer, quartz thermometer, resistance temperature detector, resistance thermometer, silicon bandgap temperature sensor, temperature gauge, thermistor, thermocouple, thermometer
Other sensors and sensor related techniques	Analog image processing, digital holography, frame grabbers, intensity sensors and their properties, atomic force microscopy, compressive sensing, hyperspectral sensors, millimeter wave scanner, magnetic resonance imaging, diffusion tensor imaging, functional magnetic resonance imaging, optical coherence tomography, positron emission tomography, quantization (signal processing), range imaging, Moire deflectometry, phase unwrapping techniques, time-of-flight camera, structured-light 3-D scanner, omnidirectional camera, catadioptric sensor, single-photon emission computed tomography (SPECT), transcranial magnetic stimulation (TMS)

different sensor sectors are on http://www.sensorsportal.com. For example, the global automotive sensor market, including silicon-based sensors, grew by 9.7 percent in 2006 to $10.1 billion and is forecast by Strategy Analytics to reach $17.1 billion by 2013 as vehicle systems such as powertrain control, safety, and convenience features become more advanced and require more sensors. IC Insights estimates that the wireless sensors and transmitters market will surpass $1.8 billion by 2012. The CMOS image sensor market alone is projected to be $8.3 billion by 2014.

Microelectromechanical systems (MEMS) is the technology of very small mechanical devices driven by electricity. It merges at the nanoscale into nanoelectromechanical systems (NEMS) and nanotechnology. MEMS are also referred to as micromachines in Japan, or microsystems technology in Europe. MEMS can be a sensor or actuator, or a transducer.

Energy harvesting (also known as power harvesting or energy scavenging) is the process by which energy is derived from external sources (e.g., solar power, thermal energy, wind energy, salinity gradients, and kinetic energy), captured, and stored for small wireless autonomous devices, like those used in wearable electronics and WSNs. Energy-harvesting devices or sensors have a very long historical connection to the water wheel, windmills, and waste heat. Before batteries (Volta, 1799) and the dynamo (Faraday, 1831), those energy-harvesting devices were the only ways to get any useful power. The following are options for energy harvesting:

- RF, used for RFID tag energy broadcasting and harvesting
- Solar, a well-known clean energy
- Thermoelectric, used in watches
- Vibrations, used in (kinetic) watches
- Human input, home utility (piezoelectric) switches

Today, there is an accelerated interest in the information and communications technology (ICT) community for powering ubiquitously deployed sensor networks, mobile electronics, electric vehicles, and so on. Many things become possible as this technolgy improves.

4.3 Connect: Via Pervasive Networks

The communications layer is the foundational infrastructure of IoT. There are two major communication technologies: wireless and wired (or wireline). Each category has broadband and narrowband, packet and circuit switched, as well as short-range and long-range communications. The penetration and traffic of U.S. wireless data subscribers in 2013 will reach the same level of broadband wired household usage in 2008 [215]. The mobile Internet is catching up quickly, thanks to the development of the Internet of Things and the flexibility of wireless communications.

Today's communications environment is a complex mix of wired and wireless networks employing circuit-switched (CS) and packet-switched (PS) technology. Developments are taking place in all four sectors and there is competition between different stakeholders, fixed mobile convergence (FMC) being an obvious example. We therefore have a communications environment that is complex [64]. We need a next-generation network (NGN), which has more than the ability to transition between circuit- and packet-switched networks. The general idea behind the NGN is that one network transports all information and services (voice, data, and all sorts of media such as video) by encapsulating these into packets, similar to those used on the Internet. NGNs are commonly built around Internet protocol, and therefore the term all-IP is also sometimes used to describe the transformation toward NGN. For example, the 3GPP long-term evolution (LTE) is a standard for wireless communication of high-speed data. It is based upon GSM/EDGE and UMTS/HSPA network technologies. One of the most important features of LTE is that it will be an all-IP flat network architecture including end-to-end QoS, provisions for low-latency communications.

With the growing abundance of embedded IoT systems comes the increased pressure at the edge of the network: multiple access methods must be accommodated, implying the need for a common underlying converged core IP/MPLS (multi-protocol label switching) network. A high-level graphic view of next-generation all-IP networking is described by Emmerson [64]. The connectivity domain enables broadband access, both wired and wireless. It also includes the transport and aggregation network. This part of the all-IP network supports various access technologies using copper lines, optical fiber, and air as transmission media.

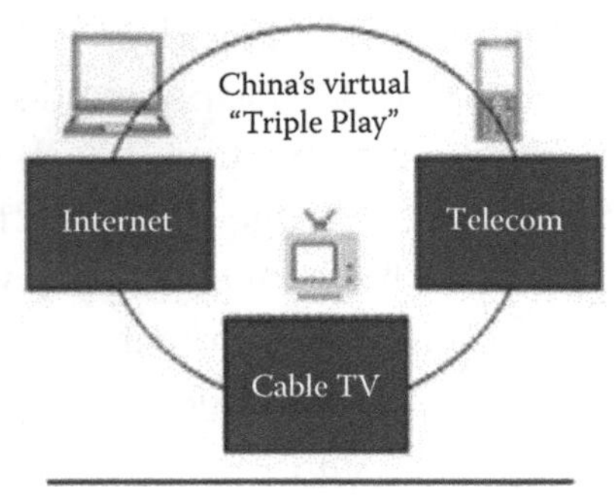

Figure 4.4 Triple network convergence.

The Chinese government has been actively pushing for the convergence of the country's three big networks—the Internet, telecom networks, and TV broadcasting networks—via various measures, most notably through the Triple Network Convergence Plan (Figure 4.4) it laid out early in 2010.

While the Triple Network Convergence Plan reiterates many government policies set out previously, one area that is expected to have significant effects on the market is the government's step to grant permission for TV broadcasting firms and telecom carriers to enter and do business in each other's realms. Local scholars estimate that triple network convergence will induce investment and consumption to 700 billion RMB (about US$103 billion), leading to widespread concern over the policy's effect on the development of related industries and various parties.

The fusion of the three networks is expected to start from business- or policy-level convergence, to application-level convergence, and finally to technological-level convergence, when the all-IP NGN vision is implemented. At that time, many good things will happen; for example, ubiquitous M2M devices can be used as cell phones, so no SIM card will be required for making a phone call.

There is no doubt that if all-IP is a reality, it will give the Internet of Things a huge lift and make the IoT dream come true much easier and faster. As an example, in the building automation industry, all-IP networking will simplify the integration work enormously, without having to deal with various field bus network protocols, OLE for process control (OPC) middleware, and so on.

Internet Protocol version 6 (IPv6) is a version of the Internet protocol that is designed to succeed Internet Protocol version 4 (IPv4). The Internet operates by transferring data in small packets that are independently routed across networks as specified by the Internet protocol. Since 1981, IPv4 has been the publicly used IP, and it is currently the foundation for most Internet communications. The Internet's growth has created a need for more addresses than IPv4 has (32 bits). IPv6 allows for vastly more numerical addresses (128 bits), but switching from IPv4 to IPv6 may be a difficult process [216].

The Internet world is getting ready for the big change from IPv4 to IPv6. After the change, everything, every duct on the planet, could have a fixed IP address, which would have an enormously huge impact on the Internet of Things on all aspects.

However, as a side note, countries such as the United States are not eager to make the change from IPv4 to IPv6 compared with countries such as China and India, because more IPv4 addresses were allocated to the United States and Europe. It's rumored that a university such as Massachusetts Institute of Technology received more IPv4 address allocation than the entire country of China or India. That's why countries such as China have developed other protocols such as IPv9 in an effort to get more IP addresses [65].

When talking about IoT, wireless communications is the topic most of the times, because three (M2M, RFID, and WSN) of the four IoT pillars are based on wireless. However, most of the systems in industrial automation, building automation, and so forth are built using SCADA technology on wired short-range field bus and long-range TCP/IP networks. The development of the Internet of Things, for the time being, should cover both wired and wireless networks, just as Axeda, the device relation management software product and service provider, did in its product and service portfolio before or after the all-IP convergence and IPv6.

4.3.1 Wired Networks

Wired networks for IoT can be categorized as short-range field bus–based access networks, mostly for SCADA applications, and IP-based networks, for M2M and SCADA applications.

The IP-based networks are widely used and their protocol stack is well known, as shown in Figure 4.5, together with telephony SS7 and cable TV DOCSIS (data-over-cable service interface specification) protocols, the triple (Internet, telephony, and cable TV) networks convergence plan candidates. SS7 (Signaling System 7) is a critical component of modern telecommunications systems (PSTN, xDSL, GPRS, etc.). Every call in every network is dependent on SS7. Likewise, every mobile phone user is dependent on SS7 to allow inter-network roaming. SS7, a form of packet switching, is also the "glue" that sticks together circuit-switched (traditional) networks with Internet protocol–based networks.

DOCSIS is an international standard that permits the addition of high-speed data transfer to an existing cable TV

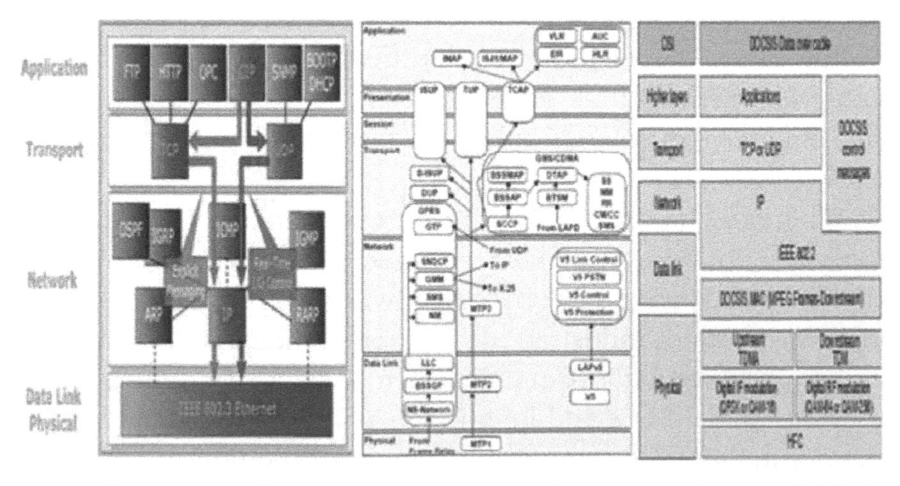

Figure 4.5 Protocol stacks of the "three networks."

system. It is employed by many cable television operators to provide Internet access over their existing HFC (hybrid fiber-coaxial) infrastructure.

A complex automated industrial system, such as a manufacturing assembly line, usually needs an organized hierarchy of controller systems to function. In this hierarchy [217,218], there is usually a SCADA/HMI (Human–Machine Interface) at the top, where an operator can monitor or operate the system. This is typically linked to a middle layer of programmable logic controllers (PLC) via a non-time-critical communications system (e.g., Ethernet). At the bottom of the control chain is the field bus (could run on top of a different power line communications network too) that links the PLCs to the IoT device components that actually do the work, such as sensors, actuators, electric motors, console lights, switches, valves, and contactors.

More details on field bus and its relevance to IoT are described here because this information is currently often neglected in most of the materials about IoT. Field bus is the name of a family of industrial computer network protocols used for real-time distributed control, now standardized as IEC 61158. The IEC 61158 standard includes eight different protocol sets called types:

- Type 1 Foundation field bus H1
- Type 2 ControlNet
- Type 3 PROFIBUS
- Type 4 P-Net
- Type 5 FOUNDATION field bus HSE (high-speed Ethernet)
- Type 6 SwiftNet (a protocol developed for Boeing, since withdrawn)
- Type 7 WorldFIP
- Type 8 Interbus

There is a wide variety of concurring standards. Table 4.2 provides a comprehensive list of wired field bus standards or protocols used with SCADA systems for industrial automation.

Table 4.2 List of Field Bus Standards

Protocol Group	*Protocols/Field Buses*
Automatic meter reading	DLMS/IEC 62056
	ANSI C12.18
	IEC 61107
	Modbus
	M-Bus
	U-SNAP [191]
Automobile/ vehicle	Local Interconnect Network (LIN)—a very low cost in-vehicle sub-network
	Controller Area Network (CAN)—an inexpensive low-speed serial bus for interconnecting automotive components
	J1939 and ISO11783—an adaptation of CAN for agricultural and commercial vehicles
	FlexRay—a general purpose high-speed protocol with safety-critical features
	Media Oriented Systems Transport (MOST)—a high-speed multimedia interface
	Keyword Protocol 2000 (KWP2000)—a protocol for automotive diagnostic devices
	Vehicle Area Network (VAN)
	DC-BUS—automotive power-line communication multiplexed network
	IDB-1394
	SMARTwireX
	J1708—RS-485 based SAE specification used in commercial vehicles, agriculture, and heavy equipment
Building, home automation	Wire—from Dallas/Maxim
	BACnet—designed by committee ASHRAE

Table 4.2 (continued) List of Field Bus Standards

Protocol Group	*Protocols/Field Buses*
	S-Bus
	C-Bus
	CC-Link Industrial Networks, supported by Mitsubishi Electric
	DALI
	DSI
	Dynet
	HomePlug—power line home networking
	HomePNA—phone line home networking
	ITU-T G.hn—a way to create a high-speed (up to 1 Gbit/s) LAN using existing home wiring (power lines, phone lines, and coaxial cables)
	Konnex (KNX)—previously AHB/EIB
	LonTalk—protocol for LonWorks by Echelon Corporation
	Modbus RTU or ASCII or TCP
	oBIX—OASIS Standard
	xAP—Open protocol
Industrial control system	MTConnect
	OPC
	OPC UA
	AS-Interface (Actuator Sensor Interface)—an industrial networking solution used in PLC, DCS, and PC-based systems
	SafetyBUS p—a standard for safe field bus communication within factory automation. It meets SIL level SIL 3 according to IEC 61508 and safety category Cat. 4 of EN 954-1

continued

Table 4.2 (continued) List of Field Bus Standards

Protocol Group	*Protocols/Field Buses*
Power system automation	IEC 61850
	IEC 60870-5
	DNP3—Distributed Network Protocol
	Modbus
	Profibus
	IEC 62351—security for IEC 60870, 61850, DNP3, and ICCP protocols
Process automation	DF-1
	FOUNDATION field bus—H1 & HSE
	Profibus—by PROFIBUS International
	PROFINET IO
	CC-Link Industrial Networks, supported by the CLPA
	CIP (Common Industrial Protocol)—can be treated as application layer common to DeviceNet, CompoNet, ControlNet and EtherNet/IP
	Controller Area Network—utilized in many network implementations, including CANopen and DeviceNet
	ControlNet—an implementation of CIP, by Allen-Bradley
	DeviceNet—an implementation of CIP, by Allen-Bradley
	DirectNet—Koyo/Automation Direct proprietary, yet documented PLC interface
	EtherNet/IP—IP stands for Industrial Protocol. An implementation of CIP, by Rockwell Automation

Table 4.2 (continued) List of Field Bus Standards

Protocol Group	*Protocols/Field Buses*
	Ethernet Powerlink—an open protocol managed by the Ethernet POWERLINK Standardization Group (EPSG)
	EtherCAT
	Interbus, Phoenix Contact's protocol for communication over serial links, now part of PROFINET IO
	HART
	Modbus RTU or ASCII or TCP
	Modbus Plus
	Modbus PEMEX
	Ethernet Global Data (EGD)—GE Fanuc PLCs (see also SRTP)
	FINS, Omron's protocol for communication over several networks, including Ethernet
	HostLink Protocol, Omron's protocol for communication over serial links
	MECHATROLINK—open protocol developed by Yaskawa
	MelsecNet, supported by Mitsubishi Electric
	Optomux—Serial (RS-422/485) network protocol originally developed by Opto 22 in 1982
	Honeywell SDS (Smart Distributed System)—originally developed by Honeywell; currently supported by Holjeron
	SERCOS interface—Open Protocol for hard real-time control of motion and I/O
	SERCOS III—Ethernet-based version of SERCOS real-time interface standard

continued

Table 4.2 (continued) List of Field Bus Standards

Protocol Group	*Protocols/Field Buses*
	GE SRTP—GE Fanuc PLCs
	Sinec H1—Siemens
	SynqNet—Danaher
	TTEthernet—TTTech
	PieP—Open Fieldbus Protocol
	BSAP—Bristol Standard Asynchronous Protocol, developed by Bristol Babcock Inc

The graphic (the CIP family of field bus protocols) in [219, first page] compares some of the field buses against the OSI model. In the past, automation field bus protocols have tended to be application specific, making them very efficient at what they do but limiting the roles for which they can be used, and making interoperability between the protocols used in different application areas difficult to achieve. The Common Industrial Protocol (CIP) forms the basis for a family of related technologies and has numerous benefits for both device manufacturers and the users of industrial automation systems. The first of the CIP-based technologies, DeviceNet, emerged in 1994 and is an implementation of CIP over CAN, which provides the data link layer for DeviceNet.

4.3.2 Wireless Networks

Just like the wired networks, wireless networks for IoT can be categorized as follows:

- Short-range (including near field communication [NFC], usually narrowband, and wireless PAN, LAN, and MAN) mesh networks, RFID, WiFi, WiMax, and so on;
- Long-range (via cellular networks, wireless WAN, pseudo-long-range) GSM, CDMA, WCDMA, and other networks, as well as satellite communication.

Short-range wireless mesh networks are the fundamental communication techniques of WSN and RFID. Long-range cellular networks are the foundation networks for M2M.

Radio spectrum refers to the part of the electromagnetic spectrum corresponding to radio frequencies: lower than 300 GHz (or wavelengths longer than about 1 mm). Different parts of the radio spectrum (as shown in http://www.ictregulationtoolkit.org/images/lib/Radio%20Spectrum%20in%20demand.gif) are used for different applications. The so-called sweet spot at ultra-high frequency concentrated most of the widely used frequencies. Radio spectrum are typically government regulated, and in some cases, are sold or licensed to operators of private radio transmission systems, for example, cellular telephone operators or broadcast television stations.

There are as many wireless standards as wired network protocols (Table 4.2). Before 2000, there were about five or six concurring standards, which lasted for a longer time than today's standard. Nowadays, there are more than 15 concurring wireless standards [220] and new ones keep coming, with each and every one's life span shorter than those before. Wireless communications standards can also be categorized as standards for cellular communications networks (such as GSM, CDMA, HSPA, LTE, etc.) and wireless connectivity networks (such as Bluetooth, Wifi, WiMax).

Communications standards are evolving rapidly. With the advent of the Internet of Things, it is expected that new standards will appear with even higher frequency and in larger numbers, due to requirements on wireless network improvements for machine-type communications (MTC) [66,189]. MTC is expected to be one of the major drivers of wireless communications standards in the next decade. The ETSI now has a technical committee exclusively focused on M2M; the Chinese Communications Standards Association is currently exploring the definition of M2M standards for China; and the Geneva-headquartered International Telecommunications Union (ITU) is working on "mobile wireless access systems providing

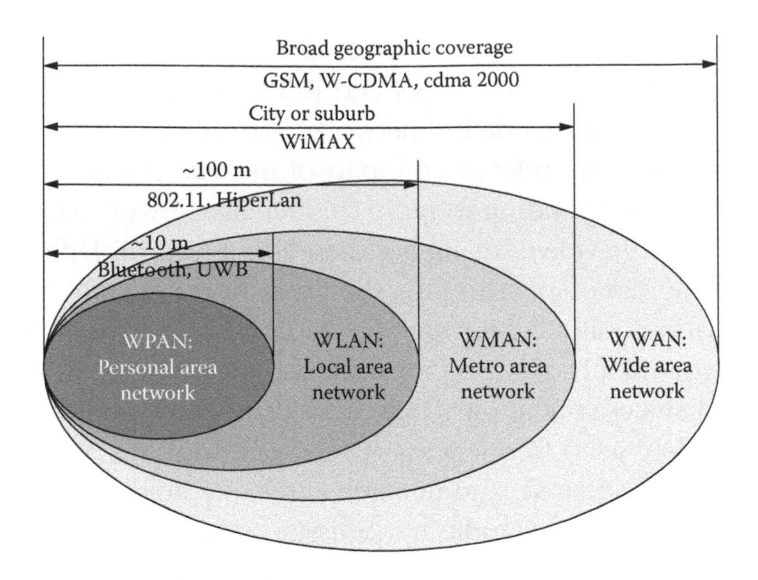

Figure 4.6 Short- and long-range wireless networks. (From Yuan Lin et al., "Baseband Processing Architectures for SDR," in Vijay Madisetti (ed.), *Wireless, Networking, Radar, Sensor Array Processing, and Nonlinear Signal Processing,* Boca Raton, FL: CRC Press, 2009.)

telecommunications for a large number of ubiquitous sensors or actuators scattered over wide areas in the land mobile service," which are at the center of the M2M ecosystem. The U.S. Telecommunications Industry Association (TIA) has also launched a new engineering committee centered on smart device communications (TIA TR-50).

Figure 4.6 shows the spectrum of wireless communications standards from short-range to long-range. RFID and NFC are parts of WPAN.

Short-range wireless sensor networks can also be treated as access networks [221] to IP-based Internet for many vertical applications such as building automation and others.

Wireless communications can be via RF, microwave (long-range line of sight via highly directional antennas, or short-range), or infrared (short-range, consumer IR devices such as remote controls). Some of the standards that have not been discussed previously are as follows:

- 6LowPAN (IPv6 over low power wireless personal area networks): a working group of IETF
- BSN (body sensor network): IEEE 802.15.6
- Broadband fixed access: LMDS, AIDAAS, HiperMAN
- DASH7: active RFID standard
- DECT (digital enhanced cordless telecommunications): cordless telephony
- EnOcean: low-power, typically battery-less, proprietary wireless technology
- HomeIR: wireless IR home networking
- HomeRF: wireless RF home networking
- IEEE 1451: a set of smart transducer interface standards by the IEEE
- InfiNET: from home automation industry leader Crestron
- INSTEON: dual-mesh technology from SmartLabs
- IrDA: from Infrared Data Association
- ISA100.11a: an open wireless networking technology standard developed by the International Society of Automation (ISA)
- Land mobile radio or professional mobile radio: TETRA, P25, OpenSky, EDACS, DMR, dPMR, etc.
- ONE-NET: open-source standard for wireless networking
- OSIAN: open-source IPv6 automation network
- TransferJet: a new type of close-proximity wireless transfer technology by touching (or bringing very close together) two electronic devices; allows high-speed exchange of data
- Wavenis: a proprietary technology by Coronis Systems in 2001. In 2008, the Wavenis Open Standard Alliance Wavenis-OSA was created to manage and govern the technology moving forward.

Apart from new standards emerging from MTC improvements, other new technologies and standards can also help in advancing the Internet of Things:

- Orthogonal frequency division multiplexing (OFDM) and orthogonal frequency division multiple access (OFDMA): two different variants of the same broadband wireless air interface. LTE is an OFDMA-based technology standardized in 3GPP. OFDM technologies typically occupy nomadic, fixed, and one-way transmission standards, ranging from TV transmission to Wi-Fi as well as fixed WiMAX and newer multicast wireless systems like Qualcomm's Forward Link Only.
- Ad hoc sensor network: a short-lived network of two or more mobile devices connected to each other without the help of intervening infrastructure. In contrast to a fixed wireless network, an ad hoc network can be deployed in remote geographical locations and requires minimal setup and administration costs. The integration of an ad hoc network with a bigger network such as the Internet or a wireless infrastructure network increases the coverage area and application domain of the ad hoc network.
- Software defined radio (SDR): SDR is the result of an evolutionary process from purely hardware-based equipment to fully software-based equipment. All functions, modes, and applications, such as transmit frequencies, modulation type, and other RF parameters, can be configured and reconfigured by software (SW) defines all waveform properties, cryptography, and applications, is reprogrammable, and may be upgraded in the field with new capabilities;
- Cognitive radio (CR): CR is a form of wireless communication in which a transceiver can intelligently detect which communication channels are in use and which are not, and instantly move into vacant channels while avoiding occupied ones. This optimizes the use of available RF spectrum while minimizing interference to other users. SDR is a required basic platform on which to build a CR. SDR and CR extend the software and middleware capabilities a

step further into the communicating devices and increase the ubiquity, versatility, and smartness of devices in the Internet of Things.

4.3.3 Satellite IoT

A communications satellite (COMSAT) is a specialized wireless transponder in space, receiving radio waves from one location and transmitting them to another (also known as a *bent pipe*). Hundreds of commercial satellites are in operation around the world. These satellites are used for such diverse purposes as wide-area network communications (to ships, vehicles, planes, as well as hand-held terminals and phones), weather forecasting, television and radio broadcasting, amateur radio communications, Internet access, and the global positioning system (GPS). Satellites have many important uses other than communications; for example, weather reports rely on satellite information, and GPS works because of a linked set of satellites. Satellite communications are especially important for transportation, aviation, maritime, and military use.

Modern communications satellites use a variety of orbits:

- GEO: Geostationary Earth Orbit, 120 satellites maximum, examples include Inmarsat (4 + 5 Satellites)
- MEO: Medium Earth Orbit, examples include the GPS satellite constellations
- LEO: Low (polar and nonpolar) Earth Orbit (theoretically unlimited); examples are Iridium (66 satellites; rent for global Iridium satellite phones is as low as $24.95 per week shown on the company's website), ORBCOMM (30 satellites), Globalstar (48), ICO (10 + 2), Ellips0 (17), Teledesic (288 satellites); constellations of satellites required for coverage
- ELI: Elliptical Orbit
- Molniya Orbit and HAPs (high-altitude platforms)

The satellite industry is a subset of the telecommunications and space industries. According to a SIA (Satellite Industry Association) report [67], the worldwide revenue of the satellite industry was $168.1 billion in 2010.

It's obvious that satellite technologies (other than positioning-oriented global navigation satellite system or GPS, which will be discussed in Chapter 6 of the book) can be used for IoT applications (such as M2M, SCADA, and telemetry) just like cellular networks, with better coverage in remote areas.

When people think of M2M communication, they usually think of cellular networks. For vehicles that move in urban areas or on major highways, cellular coverage is usually good enough, but what about construction equipment at remote locations, agricultural equipment, or ships? That's where satellite communication comes into play.

There are two issues about satellite communications: speed and cost. Although satellites can transmit large amounts of data, like Direct TV, this is done primarily one-way and to a large antenna. Two-way transmission to a small antenna has a much lower bandwidth capability than cellular communication. And the cost per byte of satellite communication is much more expensive. But if you have an application with small data requirements and broad coverage needs, satellite pricing can be very competitive.

Another option is dual-mode devices. This combines satellite and cellular in a single-edge device, giving you the best of both worlds. For example, the Axeda SmartLink platform is designed to handle dual-mode communication (via partnership with ORBCOMM) that switches between the communication modes based on price and urgency. Basic status information can be saved locally and then sent when a cellular connection is available, but an emergency condition could be sent immediately by the most economic means available.

ORBCOMM Inc. also provides M2M services to customers such as Caterpillar and Volvo Trucking, in industries ranging from commercial transportation to heavy equipment, industrial fixed assets, and marine/homeland security to track, monitor, and control their mobile and fixed assets. With ORBCOMM, these companies monitor everything from trucks and railcars to marine vessels, which are often in areas beyond the reach of terrestrial systems. ORBCOMM has deployed an M2M services portal that relies on Sierra Wireless AirVantage™ Services Platform and gives ORBCOMM's customers the ability to seamlessly track and manage their equipment around the world, even over the ocean.

In a report titled "World Satellite Machine-to-Machine Communications Market," Frost and Sullivan finds that the market earned revenues of $726 million in 2009. It estimates that this number will reach $1.90 billion in 2016. The United States dominates the world satellite M2M communications market with 62 percent market share. The Asia-Pacific region is expected to experience maximum growth in the long term. Major satellite market participants include Iridium, Inmarsat Standard C, Satamatics, SkyWave, Globalstar, Qualcomm Omnitracs, ORBCOMM, Skybitz, Wireless Matrix, and Thuraya.

Network Innovations provides a suite of mobile and stationary satellite communications solutions for wireless SCADA/telemetry data communications that operate globally. Dedicated, satellite-based business communications using relatively small dish antennas or very small aperture terminals (VSATs) are no longer only for governments and colossal corporations. An industry study predicted in 1990 that SCADA services would become a market for VSAT technology by the mid-1990s, which did happen. VSAT SCADA is now an important tool in the oil, pipeline, and electric utility markets, fulfilling the prediction. The global SCADA/M2M/LDR (low data rate) market is projected to reach US$3 billion by 2018 based on data published in April 2011 by Northern Sky Research, LLC. An example satellite SCADA system for monitoring and controlling a city's fresh water supply system is available [222].

4.4 Manage: To Create New Business Value

The previously described first two stages of the DCM model show the processes and venues of how the information is captured from various types of devices and how this information is aggregated via various gateways and transported across access networks and the core backbone to the central servers. The machine-generated information comes in large volumes much bigger and faster than information generated by humans; however, much of the data are of low value or even noises, which must be filtered out by middleware at the edge as described before in the RFID sections. And then those preprocessed data are transformed into high-value information via a cognitive application platform, most of the times a high-performance cloud computing (or high-throughput computing) platform.

In the current customer-driven, technology-based environment, it is no longer enough to offer a service or product and expect it to satisfy your customers. Even if you have the best customer service in the industry, you have to be able to extend out your offerings to meet current demand to keep the customers satisfied. The Internet of Things brings enormous possibilities and potentials for creating new business value and generating new revenue ecosystems with data processing and managing rules that combine intelligence from remote assets unreachable before with your intelligent enterprise systems.

With IoT, more and more areas of the real world become part of the ICT world, as shown in http://consen.org/node/9 from the IoSS (Internet Architecture for Optimization Sensing Systems) project in Europe. Disruptive applications beyond current imagination will appear. Smart grid, connected car, fleet control, mobile surveillance, and remote monitoring are listed as the top five disruptive applications out of a total of 65 identified, according to reports from the Boston Consulting Group. All of the top five are IoT applications. For example, with the wide use of telematics, things like total vehicle life cycle management, refined used car price estimate, Pay as

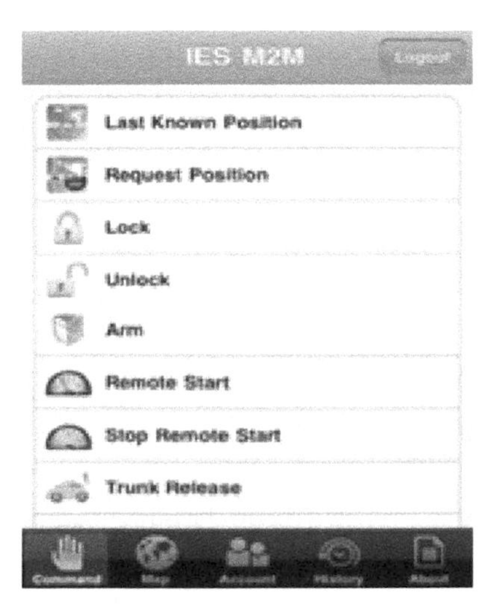

Figure 4.7 iPhone M2M application.

You Drive insurance policy, neighbor-to-neighbor car-sharing business such as those provided by startup RelayRides become possible, and the list goes on and on.

Let's take a look again at the typical capabilities of an M2M platform and how they support the business of a mobile operator or an M2M enabler/partner. With those functions and roles (as shown in http://machine2twomachine.files.wordpress.com/2011/08/fig-16.jpg [265]), both the mobile operator and the M2M partner can attain additional revenue by offering advanced services to their M2M partners (Figure 4.7). For example, the M2M platform and the fleet management system the author's team built for China Mobile utilize its existing Operation Support System/Business Support System (OSS/BSS) for SIM card issuing, billing, and other services, and China Mobile collects the revenue from the customers and shares it with us. China Unicom has also built and operates a telematics service support platform on top of their OSS/BSS, aiming to provide foundation services to a variety of TSPs (telematics service providers).

M2M applications that can be linked inside the network to people's existing mobile subscriptions offer mobile operators enormous advantages in the competitive M2M marketplace. Using smartphones as connected portable navigation devices is such an example of potentially great market growth opportunity. The application stores' model of Apple and Google Android has turned smartphones into M2M terminals. One example is the application from Portman Electronics Ltd.'s IES iPhone M2M Tracking System. It is a real time GPS/GSM/GPRS tracking service. Another example of nonoperator vendor is SeeControl, who empowers you to use sensors, GPS trackers, barcode scanners, RFID, and smart web forms to collect asset data from anywhere and manage business processes.

In the industrial automation scenario, the layering of the value chain components or subsystems looks as depicted in Figure 4.8. One of the major issues has been or still is that

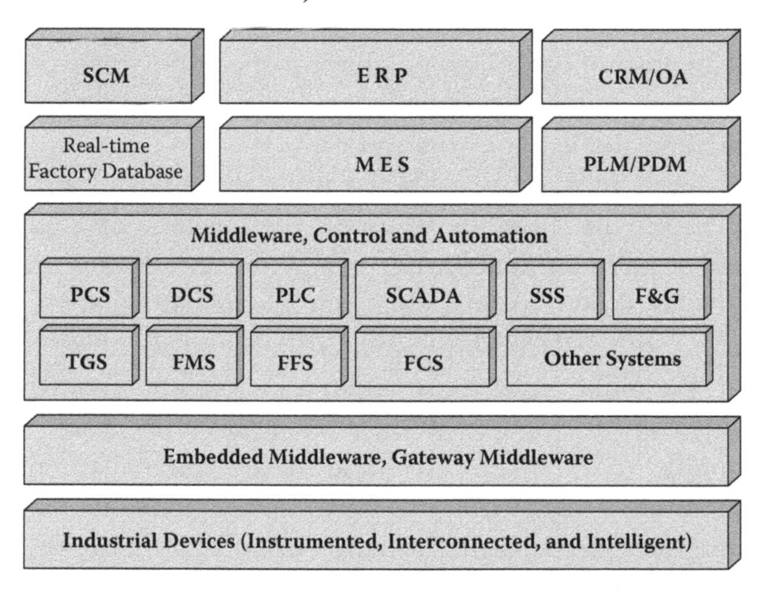

Figure 4.8 The industrial automation stack. FCS = Field Bus Control System; DCS = Distributed Control System; PLC = Programmable Logic Controller; SCADA = Supervisory Control and Data Acquisition; TMS = Tank Management System; FMS = Flow Metering System; F&G = Fire and Gas; SSS = Safety Shutdown System; FFS = Firefighting System; MES = Manufacturing Execution System; ERP = Enterprise Resource Planning.

most of these subsystems are not integrated; operators have to deal with various subsystem interfaces to run the operation. Sometimes, the factory database has to be manually keyed in to the IT database. When you want to expand and/or integrate the entire plant, you will need a solution provider with the expertise to provide the solution for you.

Before the advent of IoT or perhaps at the same time, people found out that efficient plant operations require the total integration of the field devices to the subsystems, then the integration of subsystems into a single centralized SCADA system that provides a single user interface or HMI. This is also where the new IoT system fits and sits. On top of this, those subsystems are further integrated into the MES and ERP as well as SCM, WMS, and other systems. All of those happen within an enterprise, it's an Intranet of Things ecosystem.

The vision of IoT augmented with advances in software technologies and methodologies such as SOA (service-oriented architecture), SaaS (software as a service), cloud computing, and others is causing a paradigm shift where devices can offer more advanced access to their functionality and business intelligence. As such, event-based information can be acquired, and then processed on-device and in-network. This capability provides new ground for approaches that can be more dynamic and highly sophisticated and that can take advantage of the available context. Cross-layer collaboration is expected to be a key issue in such a highly dynamic and heterogeneous infrastructure such as the Real World Internet (RWI) or IoT [68]. Device relation management and intelligent device management are some of those cross-layer M2M paradigms or product concepts proposed by Axeda and Questra a few years ago, and now those products and services are serving more and more customers.

As mentioned before, the three layers of DCM are not the run-time architecture of an IoT system, but a gross classification of the IoT value chain. For an MNO or network operator in general, IoT system architecture consists of the following seven layers [70], and the focus is on network infrastructure and service capabilities similar to those provided by telcos' existing Business and Operations Support System (BOSS).

1. M2M applications
2. M2M service capabilities
3. Core network
4. Access network
5. M2M gateway
6. M2M LAN
7. M2M devices

For other parties in the IoT value chain, the diagrams from ETSI and Digi International [69] demonstrated more generic IoT system architectures. The key is to have a single common platform that can be used for all kinds of vertical applications of IoT. The data-collecting layer of IoT, from the last mile WSNs and the gateway, to the access networks, and finally to the core network, can be distributed and replicated (and, of course, there may be cross-layer connections and Intranet of Things systems which are treated as subsystems). However, the layers above the core network should be highly integrated and centralized on top of a single common (platform as a service) PaaS + SaaS platform agnostic of and accommodating the variations of the connectivity including the IaaS (infrastructure as a service) layers.

The discussion of the PaaS + SaaS middleware layer is the focus of this book, which will be covered in more detail in the followed chapters.

In China, companies such as Datang, ZTE, and Huawei have also done extensive research on IoT/M2M because the Internet of Things is highly visible in the Chinese government and many grants have been allocated to sponsor such research activities. The sample architecture diagrams in Figure 4.9 are from Datang and ZTE.

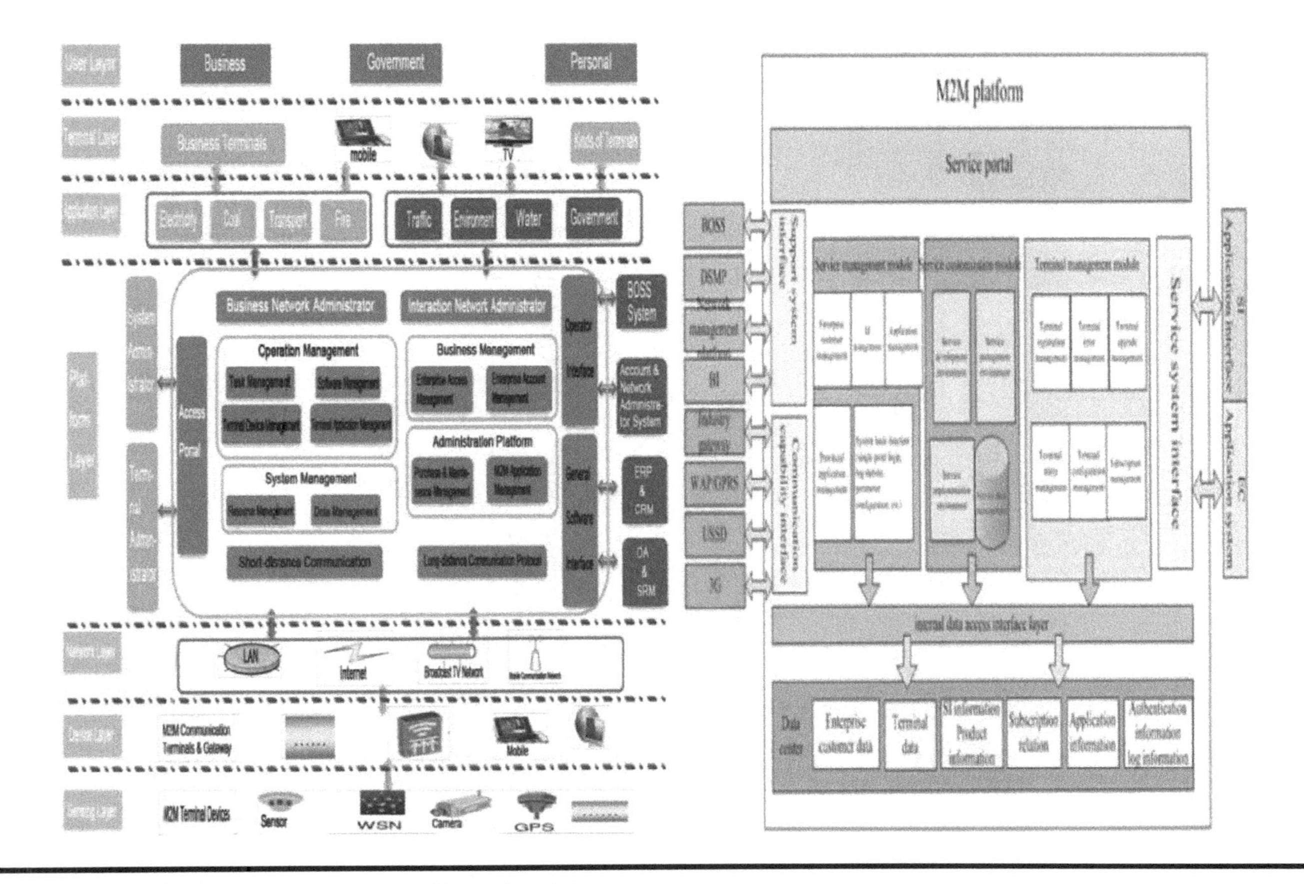

Figure 4.9 Unified IoT architecture efforts in China.

4.4.1 More Ingredients: LBS, GNSS, RTLS, and Others

Other technologies and components are widely used and required in IoT applications; however, those ingredients are not needed for all IoT systems at all times. According to SRI Consulting Business Intelligence, the technologies of the Internet of Things are summarized in Table 4.3. In the Building Blocks column we have discussed almost all of the IoT technologies, which is the goal of this book, except the Location Technology, which must be covered. Positioning capabilities and location-based services (LBS) are required for all mobility IoT applications such as telematics, fleet management, assets tracking in supply chain, and so on.

Table 4.3 IoT Technologies

Enabling Building Blocks	*Synergistic Technologies*
These technologies directly contribute to the development of the IoT.	*These technologies may add value to the IoT.*
Machine-to-machine interfaces and protocols of electronic communication	Geotagging/geocaching
Microcontrollers	Biometrics
Wireless communication	Machine vision
RFID technology	Robotics
Energy-harvesting technologies	Augmented reality
Sensors	Mirror worlds
Actuators	Telepresence and adjustable autonomy
Location technology	Life recorders and personal black boxes
Software	Tangible user interfaces
	Clean technologies

LBS is a type of context-aware computing, a term first introduced by Schilit in 1994 [71]. In 1996, the Federal Communications Commission issued the order for enhanced-911 (E-911) to provide the location of wireless callers using 911 emergency services, resulting in significant development in wireless location technologies and later location-based services. LBSs enable a customer to see the location of its devices in real time and retrieve basic information such as whether the device is registered as well as the history of data sessions. This valuable information enables the customer and the M2M solution providers to determine if the device is functioning as intended and its exact location. Should a service call, such as a part change, become necessary, they have the means to quickly and accurately locate the device. LBS can enhance the stickiness of any M2M/IoT application, especially for highly mobile solutions; new business lines and incremental revenue streams can be realized using LBS [182] creatively. A group of startups such as FourSquare, Gowalla, Loopt, myTown, BrightKite, Rummble, and others as well as Google's Latitude are providing innovative LBS services.

LBSs work using one or more of a combination of three technology protocols to determine a device's location. If the device has a GPS chip and line of sight to the navigation satellites, GPS provides the most accurate location: 15 to 100 feet. Should a pure GPS reckoning not be available due to atmospheric conditions or line-of-sight issues, assisted GPS or differential GPS will be used, providing a hybrid of satellite and cell tower location–based information, resulting in accuracy of 15 to 50 feet. If a device does not have any type of GPS technology, then enhanced cell ID will be used, which will triangulate the location of the device according to the nearest cell towers and the relative signal strength between them. This method has an accuracy of 500 to 800 feet, although it can be less accurate in more rural areas where fewer cell towers exist. More information on the major locating technologies, their accuracies, and their implementation cost range can be found in McBeath [223].

A global navigation satellite system (GNSS) is a system of satellites that provides autonomous geospatial positioning with global coverage. It allows small electronic receivers to determine their location (longitude, latitude, and altitude) to within a few meters using time signals transmitted along a line of sight by radio from satellites. Such satellites are often medium earth orbit communications satellites (discussed in the last section) that are also used for M2M communications.

The U.S. NAVSTAR GPS was the only fully operational GNSS before October 2011. The Russian GLONASS (Global Orbiting Navigation Satellite System) achieved full global coverage in October 2011 after the successful launch of the latest GLONASS satellite. China is in the process of expanding its regional Compass (Beidou) navigation system into a GNSS by 2020. The European Union's Galileo positioning system is a GNSS in initial deployment phase, scheduled to be fully operational by 2020 at the earliest. All of those GNSS satellites use CDMA for communications. The Indian Regional Navigational Satellite System is an autonomous regional satellite navigation system being developed by Indian Space Research Organization. Other countries such as France and Japan are also developing their own GNSSs.

A local positioning system (LPS) is a navigation system that provides location information in all weather, anywhere within the coverage of the network where there is an unobstructed line of sight to three or more signaling beacons of which the exact position on Earth is known. Beacons include cellular base stations, Wi-Fi access points, RFID readers, radio broadcast towers, and so on. In the past, long-range LPSs have been used for navigation of ships and aircraft. Examples are the Decca Navigator System and LORAN. Nowadays, LPSs are often used as complementary or alternative positioning technology to GPS, especially in areas where GPS does not reach or is weak, for example, inside buildings or urban canyons.

A special type of LPS is the real-time locating system (RTLS), which uses simple, inexpensive badges or tags attached to the objects, and readers receive wireless signals from these tags to determine their locations. According to IDTechEx, the market for RTLS is $380 million in 2011 rising to $1.6 billion in 2021.

A wide variety of wireless systems can be leveraged to provide real-time locating including active RFID, infrared, low-frequency signpost identification, ultrasonic ranging, ultra-wideband (UWB), Wi-Fi, Bluetooth, and so on. The locating methods or algorithms include angle of arrival, line of sight, time of arrival, time difference of arrival, time of flight, received channel power indicator, received signal strength indication, symmetrical double sided–two way ranging, near-field electromagnetic ranging; and so on.

A geographic information system (GIS)—a fusion of cartography, photogrammtery (the author worked at the Institute of Photogrammetry of ETH Zurich on related research in the late 1980s), statistical analysis, and database technology—is a system designed to capture, store, manipulate, analyze, manage, and present all types of geographically referenced data. A GIS map labeled with a variety of points of interests is a fundamental tool for many vertical IoT applications. Traditionally, maps are made up only of the more permanent fixtures of the earth's surface: roads, rivers, mountains, streets, to name a few. Over the past two decades, however, the widespread availability of GPS and mapping software has changed the landscape. Today, for example, a GPS device fed by sensors can show the state of congestion of the roads in real time on a GIS, such as the INRIX traffic services, an air-traffic controller is able to see a real-time GIS map of airplane traffic, and so on. All these possibilities and more are shifting GIS from the relatively leisurely process of analyzing static data to a far more dynamic process of real-time monitoring and decision making. With the advent of the Internet of Things, GIS will involve much more real-time situation monitoring and assessment that treat information as continually changing.

4.5 Summary

In this chapter, we talked about the technological aspects of the DCM layers of the IoT value chain. The focus was on IoT-related hardware and networks. A comprehensive collection of sensors and related technologies was discussed. A greater, more detailed overview of numerous wired and wireless, short-range and long-distance communication technologies and their mapping and relevance to and enhancements (such as MTC) for the four pillar applications of IoT were provided.

A diagram from Wireless Technologies [224] depicts the participating entities of the IoT/M2M value chain.

1. The business or consumer is involved in the consumption of the service. One possible way of their influencing the IoT is in terms of the demand. Changes in demand would lead to different configurations among the players in the business, in order to generate economically viable business models.
2. The system or service operator provides the basic M2M service to the end-user. The system operator works in tandem with the network operator to provide M2M services. The service operator has a direct relationship with the end-user.
3. The network operator provides the basic communications transport network service to the service operator.
4. The application provider or developer develops M2M value-added services for a service operator to be consumed by the end-user.
5. The end-user equipment vendor provides M2M-enabled equipment. A player in this role would typically work with the systems integrator.
6. The mobile equipment vendor provides the necessary mobile infrastructure such as GSM/GPRS/3G routers for M2M communications. A player in this role would work with the network operator.

7. The system integrator plays a major role in providing an end-to-end M2M solution. A player in this role can be an application developer and would work with network operators, end-user, and equipment vendors.

System integrators and service operators as well as application developers are in the "M" domain, network operators and equipment vendors are in the "C" domain, and end-user equipment vendors are in the "D" domain.

In the next chapter, we will be getting to the core parts of the book and talking about middleware in general and, more importantly, its role in and relevance to IoT applications.

MIDDLEWARE FOR IoT

Chapter 5

Middleware and IoT

5.1 An Overview of Middleware

There are several historical stories that linguistically unite humanity across the planet: the Tower of Babel, Enmerkar and the Lord of Aratta, Xelhua, and Toltecs. Middleware deals with the babble between distributed systems and has a similar objective in bringing linguistic or communicative unity to disparate technological systems.

The term *middleware* stems from distributed computing and refers to a set of enabling services such as standardized APIs, protocols, and infrastructure services for supporting the rapid and convenient development of distributed services and applications based on the client/server and later multitiered paradigm, which was essential for migrating single-tiered mainframe/terminal applications to multitiered architecture. Middleware is about integration and interoperability of applications and services running on heterogeneous computing and communications devices.

The services it provides, including identification, authentication, authorization, soft-switching, certification, and security,

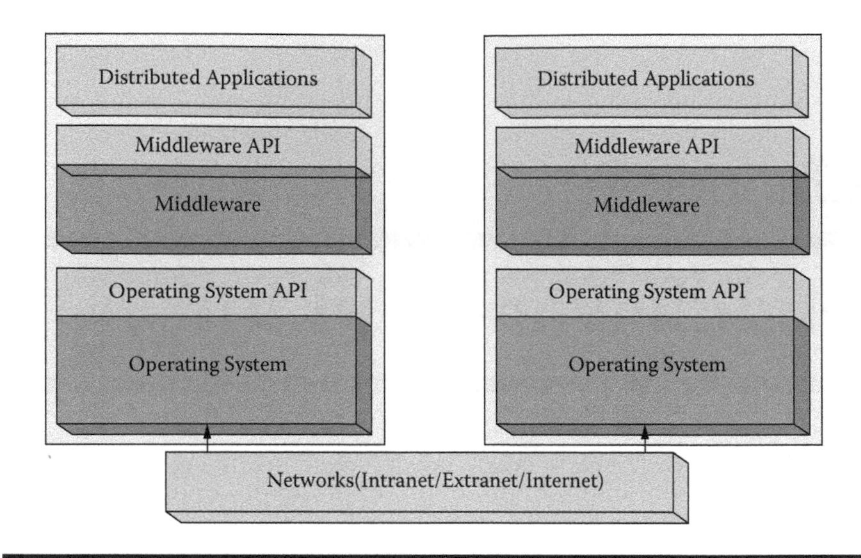

Figure 5.1 Omnipresent middleware.

are used in a vast range of global appliances and systems, from smart cards and wireless devices to mobile services and e-commerce. When the first distributed applications became widely used in the early 1990s, application developers were increasingly faced with a multitude of heterogeneous programming languages, hardware platforms, operating systems, and communication protocols, which complicated both the programming and deployment of distributed applications.

The term *middleware* refers to a layer that is arranged on top of operating systems and communications stacks and thus hides heterogeneity from the applications through a set of common, well-defined interfaces (Figure 5.1). In this way, the distributed client and server components of which an application is made up can be programmed in the same manner as if they were executed on the same host.

Middleware brings the following values to the table:

- Enables applications running across multiple platforms to communicate with each other
- Shields the developer from dependencies on network protocols, operating systems, and hardware platforms

- Is a software layer that lies between the operating system and the applications on each site of the system
- Hides heterogeneity and location independence
- Increases software portability
- Provides common functionality needed by many applications
- Aids application interoperability
- Aids scalability
- Helps integrate legacy facilities

Middleware is omnipresent and it exists nearly everywhere in an information and communications technology (ICT) system. Many kinds of middleware are described in related books [160,161]. A list of middleware is compiled below:

- Message-Oriented Middleware (MOM/MQ/JMS/ESB)
- CEP (complex event processing) Middleware (Tibco, Sybase)
- Adaptive and Reflective Middleware (TAO/DynamicTAO/OpenORB [80])
- Transaction Middleware (TPM/Tuxedo)
- Peer-to-Peer Middleware (JXTA)
- Grid Middleware (PVM/MPI/Schedulers)
- Model-Driven Middleware (CoSMIC)
- Games Middleware (Autodesk)
- Mobile Computing Middleware (OSA/Parlay/JAIN/OMA)
- Radio-frequency Identification (RFID) (Smart Cards) Middleware (Edgeware)
- Three-tiered Application Server Middleware (Weblogic, Websphere)
- Real-time CORBA Middleware (Real-time CORBA)
- High-Availability (Fault Tolerance) Middleware (Fault-Tolerant CORBA)
- Security Middleware (Siteminder)
- CATV/IPTV Middleware (MHP/GEM/OCAP) [181]
- RFID Edge Middleware (OATSystems, Sybase, Oracle, Tibco, SeeBeyond, IBM, SAP, Connectera, GlobeRanger, Manhattan Associates)

- Process-Oriented Middleware (WebMethods, SeeBeyond, Tibco, IBM, SAP, Oracle)
- Business-to-Business (B2B)-Oriented Middleware (SeeBeyond/Oracle, Tibco, webMethods)
- Middleware for Location-Based Services
- Surveillance Middleware

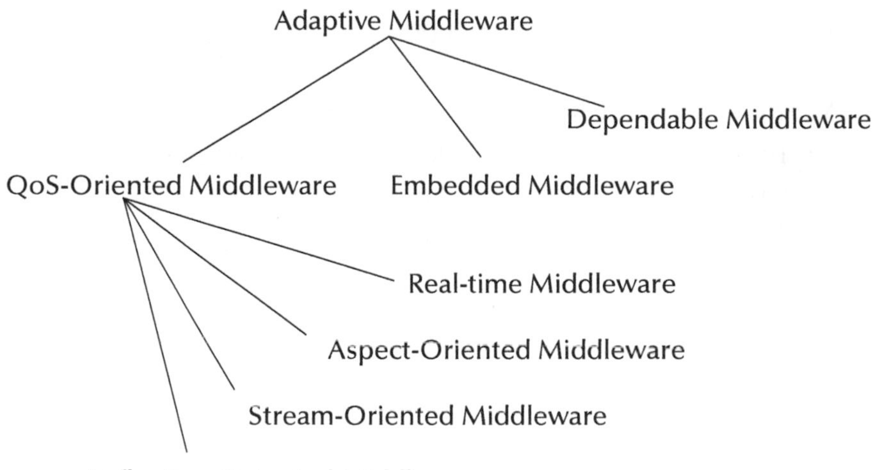

It's argued by some that with middleware proliferation these days, middleware is everywhere (there is also the concept of Everyware [19], an IoT software platform based on OSGi). It seems like every time that more than two applications need to be integrated, a piece of middleware has been deployed to handle the task. The trouble is that this has led to a lot of middleware sprawl because most of these middleware deployments are tactical, as opposed to being part of information technology (IT) strategy.

Middleware is also the software "glue" that helps programs and databases running on different computers to work together. Gartner formally defines middleware as: "Runtime system software that directly enables application-level interactions among programs in a distributed computing environment" [73].

The basis for nearly all middleware approaches was formalized by the International Organization for Standardization (ISO), which defined the common principles and structures of middleware in a framework known as Reference Model for Open Distributed Processing (RM-ODP). The main objective of ODP is to achieve distribution, interworking, and portability in an environment of heterogeneous IT resources and multiple organizational domains of different participants. ODP groups the functions of middleware into different transparency mechanisms, such as location, failure, persistence, transaction, and scalability. Each of them provides a number of APIs and services to the developer for masking the complexity associated with the respective functions.

The common principles of ODP have been adopted by many of the major middleware platforms, such as OSF DCE (Open Software Foundation's Distributed Computing Environment), Common Object Request Broker Architecture (CORBA), Java's Remote Method Invocation (RMI) and Java EE, .NET/DCOM of Microsoft, LAMP (Linux, Apache, MySQL, PHP/Perl/Python), and several approaches for web services. All of these provide several infrastructure services and support different communication patterns, for example, synchronous and asynchronous interactions.

A taxonomy of middleware functionality is outlined by Gartner [73] with three major categories: the integration middleware, the basic middleware, and the development and management tools. More than a dozen different functions that can be performed by middleware have been identified.

The integration middleware covers business- and application-oriented commonalities that include the following:

- Business process management
- Business rule engine/workflow
- Business event management
- Data routing and adapters

The basic middleware is the foundation, which applies to the Internet of Things (IoT) infrastructure also, and it can be further categorized as follows:

- Data management middleware: helps programs read from and write to remote databases or files. Examples of this kind of middleware include distributed and parallel file systems, such as Google File System, IBM GPFS, Network File System, and Windows, and also include the remote database access middleware, such as Open Database Connectivity or Java Database Connectivity libraries that are bundled into DBMSs such as IBM DB2, Oracle, and Microsoft SQL Server.
- Communication middleware: software that support protocols for transmitting messages or data between two points as well as a system programming interface (SPI) to invoke the communication service. More-advanced communication middleware (such as message-oriented middleware) also support safe (e.g., using strong security) and reliable (e.g., guaranteed once and only once) delivery of messages. Protocols and SPIs used in communication middleware can be proprietary (e.g., IBM WebSphere MQ/MQ-TT or Microsoft MSMQ) or based on industry standards such as ASN.1, DCE remote procedure call (RPC), CORBA/IIOP, Java Message Service (JMS), or web services (based on SOAP or REST). Today's communication middleware generally runs on Internet-based protocols such as HTTP (HTTPS), IP, SMTP, and so forth. It may implement higher level protocols, including industry standards (e.g., ebXML messaging and web services), and proprietary protocols (e.g., Oracle AQ), and it may run over the Internet or private networks. Communication middleware also includes *embedded middleware*. Research has been done on middleware and associated standard protocols for home automation and building controls [225,266]. Table 5.1 is a list of some emerging IoT middleware projects.

Table 5.1 Embedded IoT Middleware

IoT Middleware	*Features of Middleware*					
	Device Management	*Interoperation*	*Platform Portability*	*Context Awareness*	*Security and Privacy*	*Protocols*
HYDRA	Yes	Yes	Yes	Yes	Yes	Agnostic
ASPIRE	Yes	No	Yes	No	No	RFID
ISMB	Yes	No	Yes	No	No	RFID, etc.
UBIWARE	Yes	No	Yes	Yes	No	RFID/WiFi, etc.
UBISOAP	Yes	Yes	Yes	No	No	RFID/WiFi, etc.
UBIROAD	Yes	Yes	Yes	Yes	Yes	RFID/WiFi, etc.
GSN	Yes	No	Yes	No	Yes	RFID/WiFi, etc.
SMEPP	Yes	No	Yes	Yes	Yes	WiFi, etc.
SOCRADES	Yes	Yes	Yes	No	Yes	RFID, etc.
SIRENA	Yes	Yes	Yes	No	Yes	RFID, etc.
WHEREX	Yes	Yes	Yes	No	No	Agnostic
MQ-TT	Yes	Yes	Yes	Yes	Yes	Agnostic
Everyware	Yes	Yes	Yes	Yes	Yes	Agnostic
^{ez}M2M	Yes	Yes	Yes	Yes	Yes	Agnostic

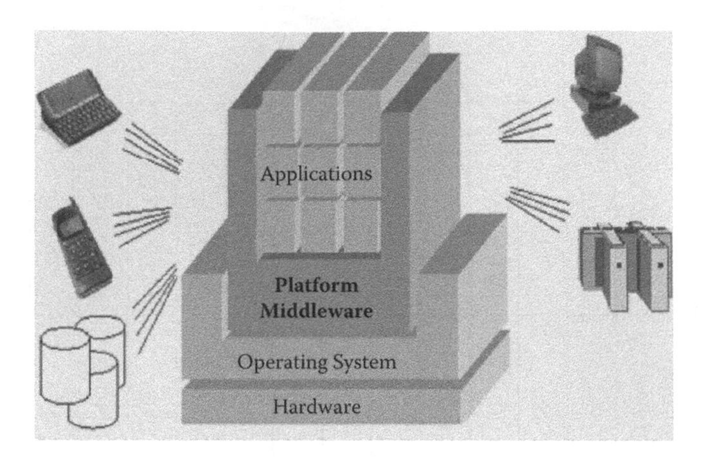

Figure 5.2 Platform middleware.

- Platform middleware: provides the runtime hosting environment (a container) for application components (see Figure 5.2). It uses embedded or external communication middleware to help programs interact with other programs. It also provides resource management services for hosting application modules at runtime (caching, starting, stopping, and multiplexing programs, load balancing, fault tolerance, access security, monitoring and management, distributed transaction processing, etc.). Platform middleware also provides interfaces to one or several forms of communication middleware (one-way messaging and request/reply). Platform middleware is well known today as *application servers* (JAVA EE or .NET Framework/COM+). However, historically, many other product categories have served as then-prevailing platform middleware. Examples include mainframe transaction processing monitors (TPMs such as IBM CICS), Unix-distributed TPMs (such as BEA Tuxedo; the author used to be part of the team), extended RPC implementations, extended object request brokers (ORBs) and object transaction monitors, DBMS stored procedures platforms, proprietary fourth-generation languages, and programmable web servers. Platform middleware has been evolving further in part because of the growing interest in *portal* services such as personalization, multichannel access, and content management. Numerous vendors offer portal services as separate products such as BEA Weblogic Portal, Plumtree, Vignette, and others that are meant to complement web servers and application servers.

Middleware and the applications software built on top of it are becoming increasingly important in the networked device marketplace. For the nonnetworked device market, the profit is from the device product itself. For the networked device market, additional profits come from consumables, services, and contents. According to Harbor Research, after the "transition point," "the device itself becomes secondary to the value it brings to the customers. Connectivity become the means to cultivate an ongoing relationship." R. Achatz, chairman at Siemens Corporate Research, noted, "We have more software developers than Oracle or SAP, but you don't see this because it is embedded in our trains, machine tools and factory automation" [171]. The landscape of CapEx (Capital Expenditure) and OpEx (Operation Expenditure) is changing.

Device miniaturization, wireless computing, and mobile communication are driving ubiquitous, pervasive, and transparent computing. Supporting these rapidly evolving technologies requires middleware solutions that address connectivity-level, location-dependent, and context-dependent issues. Many companies have developed common application *platform middleware* frameworks for M2M or IoT applications, which will be discussed in more detail in Chapter 7. We will talk more about *communication middleware* in the following sections with regard to its association with M2M or IoT applications.

5.2 Communication Middleware for IoT

In a runtime environment, the DCM (device, connect, and manage) three-layer model can be further extended into more layers depending upon the geographical scope of the area network (AN) from BAN to interplanetary Internet as listed below:

- Body (BAN)
- Personal (PAN)
- Near-me (NAN)
- Machine-to-machine, or M2M (MAN)
- Local (LAN)
  - Home (HAN)
  - Storage (SAN)
- Campus (CAN)
- Backbone
- Metropolitan (MAN)
- Wide (WAN)
- Internet
- Interplanetary Internet

In this section, we will talk about the extensions and enhancements of the existing technologies in the device and connect layers. If the IoT applications are to be extended from the current insolated Intranet or Extranet environments to the wide area as well as global Internet landscape, some fundamental changes in the networking systems have to be considered in a converged next-generation network (NGN) setting.

Some efforts such as the (open-source) Hydra project are under way to build a *unified communication network middleware* for IoT applications. Hydra [133] (networked embedded system middleware for heterogeneous physical devices in a distributed architecture) is a European Union–sponsored IoT open-source project (FP6 IST-2005-034891) that aims to reduce the complexity by developing service-oriented middleware.

5.2.1 MTC/M2M Middleware

The 3GPP (Third Generation Partnership Project) is a collaboration between groups of telecommunications associations known as the Organizational Partners. The Organizational Partners are the European Telecommunications Standards Institute (ETSI), Association of Radio Industries and Businesses/Telecommunication Technology Committee (Japan), China Communications Standards Association, Alliance for Telecommunications Industry Solutions (North America), and Telecommunications Technology Association (South Korea). The project was established in December 1998.

The connect layer of DCM can be further divided into three layers based on 3GPP's efforts for GSM/WCDMA family (3GPP2 for CDMA family) cellular wireless M2M standardization: the M2M area network layer, the access/core network layer, and the external/Internet network layer, as depicted in the 3GPP/ETSI graphic in [230]. The M2M platform in the graphic is an IoT platform middleware at the "M" layer in the DCM value chain.

- M2M area network—provide wired or wireless connectivity between M2M devices and M2M gateways, such as personal area network
- M2M access/core network—ensure M2M devices interconnection from the gateways to the access/core communication network, such as GPRS/GSM (GGSN [Gateway GPRS Support Node], SGSN [Serving GPRS Support Node], etc.; WCDMA, and others
- External/Internet networks (long distance)—communicate between the 3GPP access/core network and the M2M middleware platform for applications, such as Internet, corporate WANs, and others

Even though 3GPP introduced the concept of the M2M area network and tries to cover RFID, wireless sensor network (WSN), and supervisory control and data acquisition (SCADA) application scenarios, it is applicable for GSM/WCDMA cellular M2M only. 3GPP's coverage/scope for the entire four-pillar IoT networking possibilities are limited. Other IoT applications, for example, SCADA, may not use cellular networks at all. Those scenarios will be discussed later.

The concept of machine-type communication (MTC) was introduced by 3GPP [76]. MTC is the term 3GPP used for cellular M2M communication. It refers to communication without (or with limited) human intervention; data are input or generated by machines instead of humans, which can be significantly faster. Most future big data growth will be in the area of M2M machine-generated data, examples of which include

- Satellite-based telemetry application-generated data
- Location data such as RFID chip readings, global positioning system (GPS) output
- Temperature and other environmental sensor readings
- Sensor readings from factories and pipelines
- Output from many kinds of medical devices, in hospitals and homes alike

In 2009, Gartner estimated that data will grow by 650 percent in the following five years. Most of the growth in data is the by-product of machine-generated data, which could also create M2M data burst to the network systems. New communication middleware will play an important role in alleviating or protecting such overloads.

Current mobile networks are optimized for human-to-human communication, not for MTC. The following are some of the characteristics of MTC summarized by 3GPP (more shown in Table 5.2):

- Time tolerant—data transfer can be delayed
- Packet switched only—network operator shall provide PS service with or without a Mobile Station International Subscriber Directory Number (MSISDN)
- Online small data transmissions—MTC devices frequently send or receive small amounts of data
- Location-specific trigger—intending to trigger MTC device in a particular area, e.g., wake up the device

Table 5.2 MTC Characteristics

Characteristics *Example Applications*	*Data Volume*	*Quality of Service*	*Amount of Signaling*	*Time Sensitivity*	*Mobility*	*Server Installed Communication*
Smart energy meters	Low	Low	Intermediate	Very low	No	Yes
Road charging	Low	Low	Low	Low	Yes	No
eCall	Very low	Very high	Very low	Very high	Yes	No
Remote maintenance	Low	Low	High	High	No	Yes
Fleet management	Low	Low	Very high	Intermediate	Yes	Yes
Photo frames	Intermediate	Low	High	Low	No	Yes
Asset tracking	Low	Low	Very high	High	Yes	Yes
Mobile payments	Intermediate	Low	High	Very high	Yes	No
Media synchronization	High	Low	High	Intermediate	Yes	Yes
Surveillance cameras	Very high	Very high	Low	Very high	No	Yes
Health monitoring	High	High	High	Very high	Yes	Yes

- Group-based MTC features—MTC device may be associated with one group
- Extra-low power consumption—improving the ability of the system to efficiently service MTC applications

3GPP started the specification for MTC in early 2010; efforts are proposed as follows [66]:

- Provide network operators with lower operational costs when offering MTC services
- Reduce the impact and effort of handling large MTC groups
- Optimize network operations to minimize impact on device battery power usage
- Stimulate new MTC applications by enabling operators to offer services tailored to MTC requirements
- Prepare for number and IP address shortages

Below are issues with current telco networks for M2M:

- 3GPP SA1 has required solutions to cater for at least two orders of magnitude more devices compared with human to human.
- Shortage of telephone numbers.
- Shortage of IPv4 addresses.
- ISMI range seems large enough for most operators.

Network agnostic middleware approaches for matching application and service requirements with available network capabilities in the telecommunication domain are abundant:

- OSA-Parlay of 3GPP, Parlay-X
- JAIN (Java APIs for integrated networks)
- Open Mobile Alliance (OMA)
- Universal Plug and Play (UPnP)
- Devices Profile for Web Services (DPWS)
- Home Audio-Video Interoperability (HAVi)
- Jini and other middleware alternatives

It seems what the 3GPP's M2M effort lacks is specifying a unified middleware framework for all MTC networks. Middleware for networks is discussed in many works [78,79]. Sahin Albayrak et al. [77] emphasized that "we firmly believe that a new middleware architecture with innovative aspects in terms of: full support along the whole path rather than at the front and backend nodes, highly service aware networks, network aware services, and intelligent coordination and cooperation capabilities is the right answer to the upcoming challenges in next generation networks."

As networks evolve today, middleware based on the aforementioned OSA/Parlay, JAIN, and others for MTC is an area that requires more investigation and integration in the near future. In addition to the MTC optimization of the cellular wireless network, other optimizations or service enablement middleware (described in Chapter 3) are discussed [226,227] and their standardizations are also needed for M2M applications. Service enablement can be built as middleware that provides reliable and efficient connectivity for adjacent industry applications and to enable operators to

- Act as horizontal service providers across applications and industries
- Expand their role as managed service providers
- Capture maximum value as smart service providers

Nokia is one of the earliest vendors that offered M2M middleware. The Nokia M2M platform [228] is based on open, widely accepted middleware (built on CORBA) and communications architecture, and it supports standard GSM technology with a choice of wireless bearers. Open interfaces facilitate easy development, operation, and maintenance of various M2M applications and services, and provide an easy upgrade path for future technologies. IBM also built an MQ-TT (telemetry transport) middleware (http://mqtt.org/) for M2M applications over IP and non-IP networks.

Other kinds of M2M terminals are the CATV STB (set top box), globally executable MHP (GEM), and MHP (multimedia home platform, based on Java technologies) [128]. These are two of the middleware standards for cable TV, IPTV, Blu-Ray player terminals (embedded middleware), and head-end (platform middleware) applications. GEM, based on MHP, is also a recommended standard by ETSI and ITU.

STB-based home gateway terminal is also an important IoT/M2M application that has been developed for many years. Other middleware for STB M2M devices and head-end systems include Multimedia and Hypermedia Information Coding Expert Group (MHEG), Open Cable Application Program (OCAP), OpenTV, MediaHighway, Digital Video Broadcasting (DVB)-HTML, etc. All-IP convergence applications based on converged middleware will make the "triple network convergence" of China a reality.

In the digital home (or home automation, domotics) scenario, middleware technology refers to a layer of software that lies on top of a home device's or appliance's operating system. Middleware facilitates rapid development and increases scalability of a system and integration of services in digital homes. It bundles hardware and software into a single solution and provides transparent interaction between home systems and databases, enables unified user interfaces, reduces infrastructure requirements, and makes multiple services easier to manage. A typical digital home could have a number of home devices and appliances, which allows the physical interconnection of multiple systems and services. Home systems and services are inevitably supplied by different manufacturers and use a wide range of different protocols and standards for communication. The home systems and services must be interconnected seamlessly with a consistent middleware platform. An example of the integration architecture of middleware with various digital home services based on standards such as UPnP, DPWS, Jini, HAVi, and so forth is available [231].

5.2.2 SCADA Middleware

The concept of MAN (M2M area network) was introduced in 3GPP/ETSI's MTC specification. This concept also applies to other pillar segments of IoT. However, not all IoT applications will use a cellular network. In fact, most of the traditional SCADA applications have been using local wireline networks for communications. The remote terminal units (RTUs), programmable logic controllers (PLCs), or even process control systems (PCSs) communicate to the SCADA middleware server via gateways (similar to MAN but all wired) that aggregate data from different wired field buses. The SCADA system is accessed in a LAN environment (sometimes xDSL, cable, WiFi, or WiMax can be used) before it is integrated into the corporate back office system.

Considering that many of the field buses also support IP, such as Modbus TCP/IP, BacNet IP, and others, it is possible or easier than wireless networks to adopt an all-IP approach to implement SCADA applications. This approach has been used in some of the projects done by the author in building management systems. Figure 5.3 (redrawn based on concepts from [264]) depicts the role of SCADA middleware in such a scenario in more detail.

Companies providing such SCADA middleware products include the following:

- Central Data Control: CDC provides the software platform Integra, which utilizes data agents to translate protocols from different building system components into single management system.
- Elutions: Its Control Maestro product has a SCADA heritage. SCADA may be best known for industrial processes but is also deployed for infrastructure (water treatment plants, gas pipelines, etc.) as well as facility systems. Control Maestro is web-based, uses human–machine

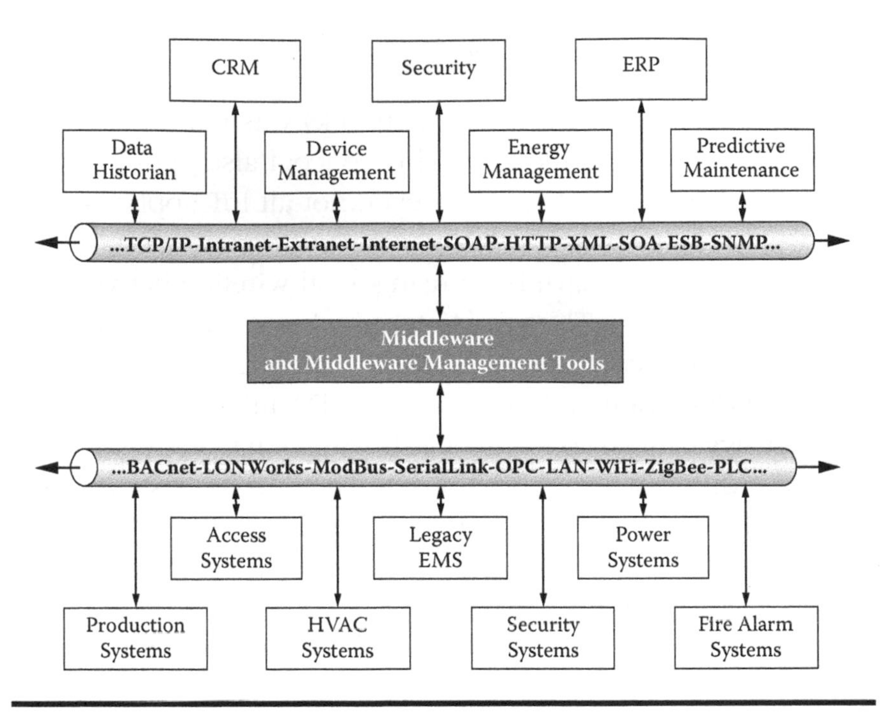

Figure 5.3 SCADA middleware architecture.

interfaces (HMI), and is able to deliver real-time and historical information.

- Richards Zeta: RZ's middleware solution is a combination of system controllers and software.
- Tridium: It provides the Niagara Java-based middleware framework and JACE hardware controllers. The Niagara platform provides protocol translation for a range of systems and the tools to build applications. Niagara has open APIs to all Niagara services and an extensible component model (XML) that enable development of applications by third parties. It also provides support for web-services data handling and communications with enterprise applications.

With the development of wireless technologies, systems have been developd that blend wireless with wired communication in SCADA applications. SensiLink™ is a middleware and software suite from MeshNetics that links wireless sensor networks with SCADA systems. Sensor data collected from the nodes is channeled through RS232, RS485, USB, Ethernet, or GPRS gateway to the SensiLink server.

OPC middleware products are one of the important communications layer SCADA middleware that are designed to enhance any OPC standards-based applications. Originally, OPC was defined as a standardized solution for the recurring task of connecting PC-based SCADA/HMI applications with automation and process control devices. Today, the OPC standard has evolved into a robust data carrier able to transport entire enterprise resource planning documents and even video signals.

OPC is for Windows only (details about the standard is discussed in Chapter 6). Tridium is arguably the first SCADA middleware based on Java technology. Recent developments have integrated new technologies such as Java and iOS (application store) to build OS platform agnostic middleware for broader IoT applications; adopting new technologies for SCADA is a trend.

5.2.3 RFID Middleware

RFID networking shares a similar three-tiered communication architecture (as shown in Figure 5.4). RFID readers are the gateways similar to MAN. Data from the readers go to the corporate LAN and then are transmitted to the Internet as needed. However, just like the scenarios of M2M and SCADA, most current RFID systems stop at the corporate LAN level and are IoT systems only.

RFID middleware (including the edge middleware or edgeware) is currently no doubt the most well-defined, comprehensive, standardized middleware compared with the other three pillar segments of IoT. Before 2004, an RFID middleware-based system was defined by EPCglobal, which included:

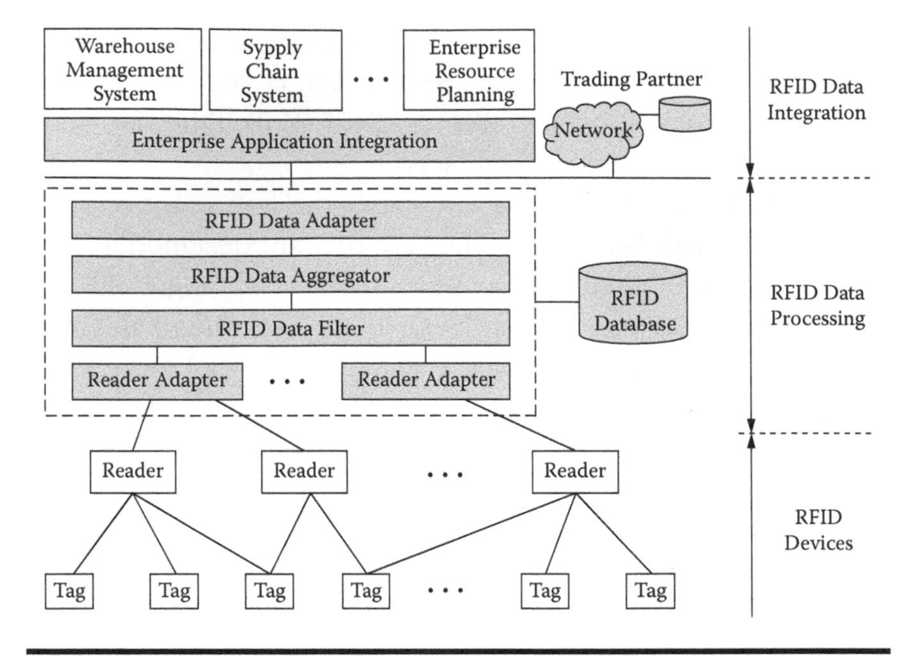

Figure 5.4 RFID architecture. (From Quan Z. Sheng, Kerry L. Taylor, Zakaria Maamar, and Paul Brebner, "RFID Data Management: Issues, Solutions, and Directions," in Lu Yan, Yan Zhang, Laurence T. Yang, and Huansheng Ning (Eds.), *The Internet of Things: From RFID to the Next-Generation Pervasive Networked Systems*, New York: Auerbach Publications, 2008.)

- A format for the data called physical markup language (PML), based on XML (Figure 5.5 is an example)
- An interface to the servers containing PML records
- A directory service called ONS (object naming service), analogous to the DNS. Given a tag's EPC, the ONS will provide pointers to the PML servers containing records related to that tag.

However, since 2004, the unified PML schema has been dropped [51] due to, most likely, practical reasons because most RFID systems are still in the "Intranet of Things" scope. Using the generic PML/ONS approach would involve overhead and sacrifice efficiency. Instead, the PML-like schema was left to the vertical applications to define their own XML

```
<pmlcore:Sensor>
    <pmluid:ID>urn:epc:1:4.16.36</pmluid:ID>
    <pmlcore:Observation>
        <pmlcore:DateTime>2002-11-06T13:04:34-06:00</pmlcore:DateTime>
        <pmlcore:Tag>
            <pmluid:ID>urn:epc:1:2.24.400</pmluid:ID>
            <pmlcore:Sensor>
                <pmluid:ID>urn:epc:1:12.8.128</pmluid:ID>
                <pmlcore:Observation>
                    <pmlcore:DateTime>2002-11-06T11:00:00-
06:00</pmlcore:DateTime>
                    <pmlcore:Data>
                        <pmlcore:XML>
                            <TemperatureReading xmlns="http://sensor.example.org/">
                                <Unit>Celsius</Unit>
                                <Value>5.3</Value>
                            </TemperatureReading>
                        </pmlcore:XML>
                    </pmlcore:Data>
                </pmlcore:Observation>
                <pmlcore:Observation>
                    <pmlcore:DateTime>2002-11-06T12:00:00-
06:00</pmlcore:DateTime>
                        ..........
                        ..........
```

Figure 5.5 Physical markup language sample.

scheme. Consequently, the overall system architecture of RFID has evolved from a dedicated structure to a more generic, open architecture.

However, the PML approach is believed to be a good IoT data representation method that should be used when the day of the full-blown IoT system comes. Other efforts such as M2MXML (from BiTX) and oBIX (an OASIS standard) are under way that are trying to build a generic IoT data schema, which is discussed in the next chapter.

An example of commercial RFID middleware product is IBM's WebSphere Sensor Events. WebSphere Sensor Events delivers new and enhanced capabilities to create a robust, flexible, and scalable platform for capturing new business value from sensor data. WebSphere Sensor Events is the platform for integrating new sensor data, identifying the relevant business events from that data using situational event processing, and then integrating and acting upon those events with SOA business processes.

The blending or convergence of different pillar IoT applications to build cross-segment IoT systems is a trend that has been demonstrated [228], in which unified data representation and associated communication middleware became more and more important.

5.2.4 WSN Middleware

Middleware also can refer to software and tools that can help hide the complexity and heterogeneity of the underlying hardware and network platforms, ease the management of system resources, and increase the stableness of application executions. WSN middleware is a kind of middleware providing the desired services for sensor-based pervasive computing applications that make use of a WSN and the related embedded operating system or firmware of the sensor nodes [57]. In most cases, WSN middleware is implemented as embedded middleware on the node [82].

It should be noted that while most existing distributed system middleware techniques aim at providing transparency abstractions by hiding the context information, WSN-based applications are usually required to be context aware, as mentioned in Chapter 1 [18].

A complete WSN middleware solution should include four major components: programming abstractions, system services, runtime support, and quality of service (QoS) mechanisms. Programming abstractions define the interface of the middleware to the application programmer. System services provide implementations to achieve the abstractions. Runtime support serves as an extension of the embedded operating system to support the middleware services. QoS mechanisms define the QoS constraints of the system. The system architecture of WSN middleware is shown in Figure 5.6.

Middleware for WSN should also facilitate development, maintenance, deployment, and execution of sensing-based applications. Many challenges arise in designing middleware for WSN due to the following reasons and more:

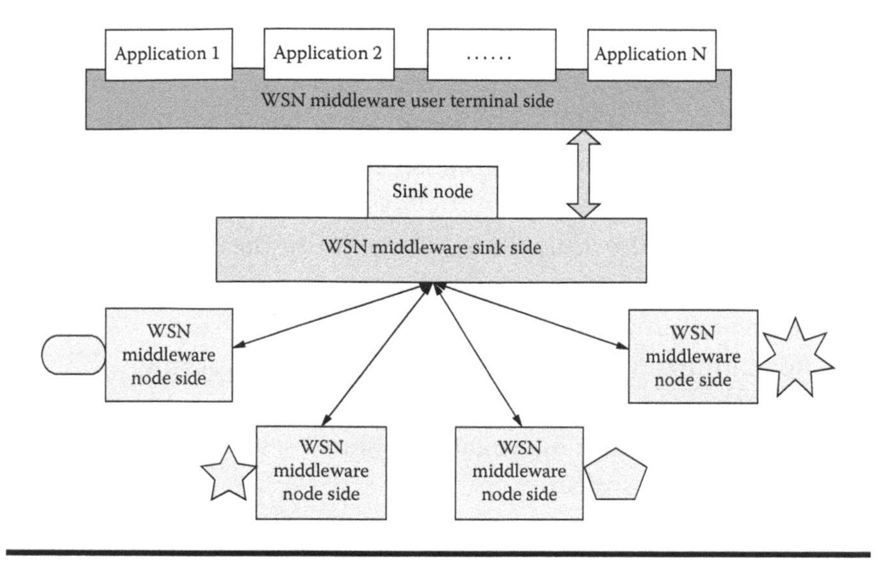

Figure 5.6 WSN middleware architecture.

- Limited power and resources, e.g., battery issues
- Mobile and dynamic network topology
- Heterogeneity, various kinds of hardware and network protocols
- Dynamic network organization, ad-hoc capability

WSN middleware is designed using a number of approaches such as virtual machine, mobile agents, database based, message-oriented, and more. Example middleware are as follows [83]:

- MagnetOS (Cornell University): power-aware, adaptive; the whole network appears as a single JVM, standard Java programs are rewritten by MAGNET as network components, and components may then be "injected" into the network using a power-optimized scheme.
- IMPALA: modular; efficiency of updates and support dynamic applications; application adaption with different profiles possible; energy efficient; used in the ZebraNet project for wildlife monitoring.

- Cougar: represents all sensors and sensor data in a relational database; control of sensors and extracting data occurs through special SQL-like queries; decentralized implementation; message passing based on controlled flooding.
- SINA (system information networking architecture): based on a spreadsheet database wherein the network is a collection of data sheets and cells are attributes; attribute-based naming; queries performed in an SQL-like language; decentralized implementation based on clustering.
- MIRES: publish/subscribe; multihop routing; additional service (e.g., data aggregation); sense—advertise over P/S and route to sink.
- MQTT-S (Message Queue Telemetry Transport for Sensors, IBM): a publish/subscribe messaging protocol for WSN, with the aim of extending the MQTT protocol beyond the reach of TCP/IP infrastructures (non-TCP/IP networks, such as Zigbee) for sensor and actuator solutions; a commercial product.
- MiLAN: provides a mechanism that allows for the adaptation of different routing protocols; sits on top of multiple physical networks; acts as a layer that allows network-specific plug-ins to convert MiLAN commands to protocol-specific ones that are passed through the usual network protocol stack; can continuously adapt to the specific features of whichever network is being used in the communication.

The WSN middleware is considered to be "proactive" middleware in the middleware family. A more comprehensive list of existing WSN middleware platforms, software/OS, and programming languages is shown in Table 5.3. A comparison of some of the WSN middleware is available [84].

As an example, the Agilla middleware is examined here in more detail (Figure 5.7). The Agilla [229] runs on top of TinyOS and allows multiple agents to execute on each node. The number of agents is variable and is determined primarily by the

Table 5.3 Sample WSN Middleware and WSN Languages

WSN Middleware			
Agilla	eCos	MagnetOS	SINA
AutoSec	EMW	MANTIS	SOS
Bertha	Enviro-Track	Mate	TinyDB
BTnut Nut/OS	EYESOS	MiLAN	TinyGALS
COMiS	FACTS	Mire	TinyOS
Contiki	Global Sensor Networks (GSN)	Netwiser	t-Kernel
CORMOS	Impala	OCTAVEX	VIP Bridge
COUGAR	jWebDust	SenOS	
DSWare	LiteOS	SensorWare	
WSN Languages			
c@t	DCL (Distributed Compositional Language)	galsC	nesC
Protothreads	SNACK	SQTL	

amount of memory available. Each agent is autonomous but shares middleware resources with other agents in the system.

Agilla provides two fundamental resources on each node: a neighbor list and a tuple space. The neighbor list contains the addresses of neighboring nodes. This is necessary for agents to decide where they want to move or clone to next. The tuple space provides an elegant decoupled-style of communication between agents. It is a shared memory architecture that is addressed by field-matching rather than memory addresses. A tuple is a sequence of typed data objects that is inserted into the tuple space. The tuple remains in the tuple space even if the agent that inserted it dies or moves away. Later, another agent may retrieve the tuple by issuing a query for a tuple with the same sequence

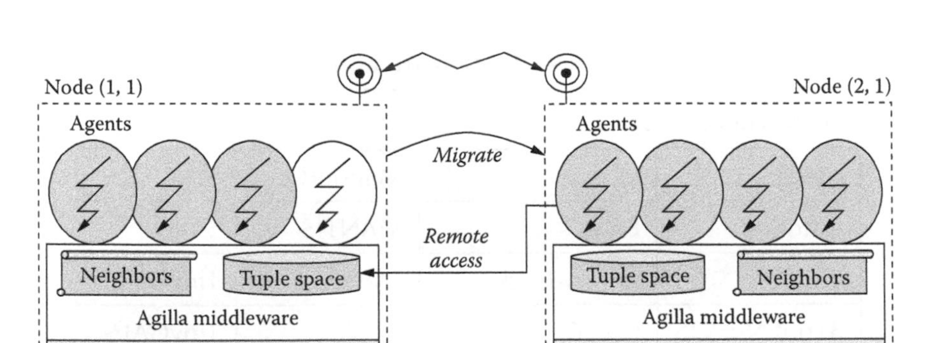

Figure 5.7 The Agilla middleware model. (From Chien-Liang Fok, Gruia-Catalin Roman, and Chenyang Lu, "Software Support for Application Development in Wireless Sensor Networks," in Paolo Bellavista and Antonio Corradi (Eds.), *The Handbook of Mobile Middleware*, New York: Auerbach Publications, 2006.)

of fields. Note that tuple spaces decouple the sending agent from the receiving agent: they do not have to be co-located, or even aware of each other's existence, for them to communicate. This is basically a fault-tolerant distributed computing technology.

All of the above WSN middleware are at the device level up to the gateways (equivalent to the MAN of MTC). Most of them are research projects conducted at universities and research institutions with a few experimental uses and of limited commercial value. This situation is very much like the research on parallel computing architecture one or two decades ago. There was a proliferation of parallel architectures [85] such as hypercube, wavefront arrays, pyramids, systolic arrays, and others, which the author has gone through [86–95]. Many research papers have been produced, but none of these architectures exist in the real world now. Nowadays, 99 percent of the world's fastest high performance computing (HPC) supercomputers use the simple massive parallel processing (MPP) architecture [96]. David Culler, the inventor of TinyOS and Mote, professor at University of California–Berkeley, was one of the prominent researchers on parallel architecture at that time. He has been doing research on WSN since the wane of parallel architecture research. In fact, some of the WSN architecture and middleware ideas are inherited from parallel computer architectures, which will most likely diminish the same way as time passes by, especially the ad hoc wireless networks (they may have greater value in military uses).

Nevertheless, once the data from the ad hoc mesh WSN reaches the gateways, or if the wireless sensors are directly connected to the higher-tier networks, the remaining process and route to reach the Internet of Things will be the same as the other pillar segments of IoT. The WSN middleware at the system level may be the same as SCADA or M2M or RFID systems, which share the same three-tiered architecture discussed in the last three sections.

5.3 LBS and Surveillance Middleware

Other than the communication middleware and the platform middleware (which will be covered in Chapter 7) for IoT applications, other middleware are related IoT or are part of IoT. Location-based service (LBS) and surveillance middleware are two of the examples we choose to cover in this chapter.

LBS is a service that integrates a mobile device's location or position with other information so as to provide added value to a user [97]. There are several uses of LBS, and some of them are direct IoT applications:

- News: information dissemination based on the location of a user, such as weather information
- Point of interest (POI): shows points of interest near the user or vehicles
- Directions: shows directions from the current location of a user
- Yellow pages: finds services near the user

- Fleet management: tracks positions of a transportation fleet
- Local advertisement: user receives advertisements according to his or her position
- Emergency: tracks current position of a user in an emergency
- Location-based games: player interacts with another player according to his or her position

LBS scenarios involve collecting, analyzing, and matching different types of information including user profiles (e.g., personal information and interests) and information dissemination profiles. For each piece of information, LBS systems have to handle different aspects:

- Spatial: LBS middleware must be able to collect information about mobile position and fixed elements, associate them with physical/logical maps, and efficiently match locations and regions.
- Temporality: Location information has a temporal dimension that must be included in query capability.
- Inaccuracy, imprecision, and uncertainty: LBS must deal with inaccuracy and imprecision associated with location positioning technologies.
- Large volumes: In real scenarios, LBS must handle large volumes of data; scalability is a very important issue.
- Continuous queries: In an LBS scenario, query executions are continuous, so the query engine of an LBS middleware must be efficient.

An example middleware architecture for LBS systems can be found at locationet.com (http://www.locationet.com/LBSmiddleware.php). Most LBS middleware can be categorized as event based (publish/subscribe), tuple space based, context aware, and data sharing based:

- Publish/subscribe: one of the most prominent middleware models, in which communication is defined in terms of exchanging asynchronous messages based on subscription.
- Tuple space: originally proposed to coordinate concurrent activities in parallel programming systems such as Linda, in which a process communicates with another process in a global collection of tuples. A tuple is a data element that contains values of a specified data type.
- DBMS-based: comprises the use of database interaction to implement a communication and coordination; many geographic information systems (GISs) operate according to this scheme. LBS architecture naturally fits the DBMS-model, such as user management systems and accounting information systems.

As an example, LocatioNet middleware is a product that meets mobile operators' needs for in-house location-privacy management, location billing functionality, provisioning interfaces, and links to various content databases. LocatioNet comprises a set of modules offered in any required combination:

- Comprehensive location privacy management: allows users to decide who can see their location, when, and how precisely, application by application
- Billing for location: gives operators a flexible set of billing options for their location and GPS services
- Provisioning: enables operators to provision user-to-location and GPS applications
- Content interfaces: enables operators to take advantage of content properties they have access to (such as local news, the weather, points of interest, traffic) by linking them to the location and GPS infrastructure

A Location API for J2ME has been specified as JSR-179 that enables mobile location-based applications for resource-limited devices. Java middleware and applications can be developed based on the Location API standard. The Open GIS Consortium (OGC) also produced a specification about location services called OpenLS™ in 2003.

Automated video surveillance networks are a class of sensor networks (people argued that a video surveillance network without automatic image recognition and event detection or alert generation is not a sensor network but instead simply a video or image capture and transmission system) with the potential to enhance the protection of facilities such as airports and power stations from a wide range of threats. However, current systems are limited to networks of tens of cameras, not the thousands required to protect major facilities. Realizing thousand-camera automated surveillance networks demands sophisticated middleware and architectural support as well as replacing the ad hoc approaches used in current systems with robust and scalable methods.

The IBM Smart Surveillance Solution [100] is based on the MILS (middleware for large-scale surveillance) surveillance middleware and designed to work with a number of video management systems from partner companies. The MILS provides the data management services needed to build a large-scale smart surveillance application. While MILS builds on the extensive capabilities of IBM's Content Manager and DB2 systems, it is essentially independent of these products and can be implemented on top of third-party relational databases. The MILS take the automatically detected events from the SSE (smart surveillance engine) as inputs. An SSE is a class of surveillance algorithms such as the HMM (Hidden Markov Model) [99].

The IBM SSS system provides two distinct functionalities:

- Real-time user-defined alerts: The user defines the criteria for alerting with reference to a specific camera view, for example, parked car detection, tripwire, and so forth.

- Indexed event search: The system automatically generates descriptions of events that occur in the scene and stores them in an indexed database to allow the user to perform a rapid search.

Another middleware approach for video surveillance networks is proposed [98]. This surveillance middleware approach partitions systems based on an activity topology—a graph describing activity observed by the surveillance camera network. Processing within topological partitions uses well-known architectural styles such as blackboards, and pipes and filters. Communication between partitions uses a service-oriented architecture. This middleware enables building intelligent video surveillance systems at a far larger scale than was previously possible. Communication on the surveillance network follows the service-oriented model with publish/subscribe messaging, providing scalability, availability, and the ability to integrate separately developed surveillance services.

5.4 Summary

Middleware is a piece of reusable software that communicates to other processes, most of the time over a network connection. This is essential for IoT applications. In this chapter, a comprehensive overview and definition of middleware and their application/relevance to IoT is presented including general purpose, horizontal multitiered platform middleware, communication middleware, embedded middleware, LBS middleware, and others. A number of vertical application-oriented middleware and their relation to the four IoT pillars are also discussed.

Middleware makes more sense if based on standards, is vertical application agnostic, and can be used as a horizontal platform serving many vertical applications. Two kinds of standardizations are important for middleware: standardized data representations and standardized architecture or frameworks. These two kinds of standards for IoT will be discussed in the next two chapters.

In the next chapter, data representations and protocols for the four IoT pillars and their unified standardization possibilities for a unified IoT middleware framework/architecture are discussed.

Chapter 6

Protocol Standardization for IoT

6.1 Web of Things versus Internet of Things

Vinton Cerf is one of the inventors of TCP/IP (transmission control protocol/Internet protocol) around 1978, which was based on his Ph.D. advisor Leonard Kleinrock's packet-switching theory published in 1961. TCP/IP became the required protocol of ARPANET (Advanced Research Projects Agency Network) in 1983. It also allowed ARPANET to expand into the Internet, facilitating features like remote login via Telnet, and later, the World Wide Web. During his tenure from 1976 to 1982 as project manager and principle scientist at DARPA (Defense Advanced Research Projects Agency), Cerf was at the center of the global network's transformation and played a key role in leading the development of the TCP/IP protocols and the Internet. Cerf is credited as father of the Internet.

Tim Berners-Lee was the man leading the development of the World Wide Web, the defining of HTML (hypertext markup language), HTTP (hypertext transfer protocol), and URL (universal resource locator), used to create web pages. All of those developments took place between 1989 and 1991. For many people who are not tech savvy, the Internet and Web are one and the same. Many people believe Tim Berners-Lee is the father of the Internet due to the success of the World Wide Web. As the Internet existed long before the World Wide Web, Tim Berners-Lee is only "old enough" to be the father of the Web.

We need to distinguish the difference between the Internet and the World Wide Web here. The Internet is the term used to identify the massive interconnection of computer networks around the world. It refers to the physical connection of the paths between two or more computers. The World Wide Web is the general name for accessing the Internet via HTTP, thus www.anything.something. It is just one of the connection protocols that is available in the Internet, and not the only one. The Internet is the large container, and the web is a part within the container. It is common in daily conversation to discuss them as the Internet and the web, and it is a very common mistake for most people to treat the Internet and web as if they were interchangeable, although it can be argued that the World Wide Web is the most popular method of using the Internet. To be technically precise, if the Internet is the restaurant, the web is the most popular dish on the menu. However, it's *the dishes* (in Figure 6.1) that make the Internet popular, useful to everyone, and powerful.

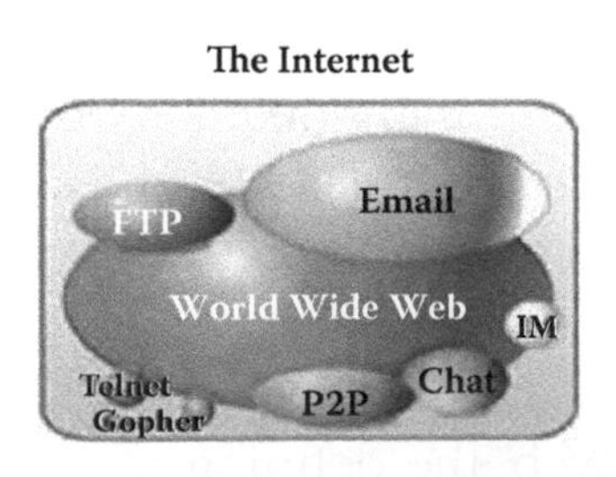

Figure 6.1 Major Internet applications.

By the same token, the key to make the Internet of Things (IoT) take off is the Web of Things (WoT)—the killer applications' platform or base of the IoT. The Web of Things is the next logical step in this IoT evolution toward global networks of sensors and actuators, enabling new applications and providing new opportunities. The Web of Things explores the layer on top of connectivity with things and addresses issues such as fast prototyping, data integration, and interaction with objects. Because the web is omnipresent and flexible enough, it has become an excellent protocol for interacting with embedded devices, and the Web of Things is a vision where things become seamlessly integrated into the web—not just through web-based user interfaces of custom applications, but by reusing the architectural principles of the web for interacting with the quickly expanding ecosystem of devices or embedded devices that are built into everyday smart objects. Well-accepted and well-understood standards and blueprints (such as uniform resource identifier [URI], HTTP, RESTful API, Atom Syndication Format) are used to access the functionality of the smart objects.

The IoT is by definition global and should be considered global in the context even that legislative and regulatory inquiries must be considered locally, regionally, nationally, and internationally. As a matter of fact, lots of IoT work has inevitably been in the WoT arena; however, it's still important to make the disctinction between IoT and WoT. One of the early prototypes mentioning the WoT concept is the Energy Visible project at ETH Zurich [101] in which sensors capable of monitoring and controlling the energy consumption of household appliances offer a RESTful API to their functionality. This API is then used to create a physical mashup. Nimbits (http://www.nimbits.com) is an open-source data historian server built on cloud computing architecture that provides connectivity between devices using data points.

There are also many other WoT applications around the world. WoT portals also started to appear just like the Internet portals (public websites) such as Yahoo, Sina, and so forth in the early days of the Internet revolution. Some of the WoT applications are listed here. More will be discussed in the next chapter.

- Arduino (http://arduino.cc/en/): Arduino can sense the environment by receiving input from a variety of sensors and can affect its surroundings by controlling lights, motors, and other actuators.
- Japan Geiger Map (http://japan.failedrobot.com/): this map visualizes crowd-sourced radiation Geiger counter readings from across Japan.
- Nanode (http://nanode.eu/): Nanode is an open-source Arduino-like board that has built-in web connectivity. It is a low-cost platform for creative development of web-connected ideas.
- The National Weather Study Project (http://nwsp.ntu.edu.sg/sensormap/): NWSP is a large-scale environmental study project deploying hundreds of mini weather stations in schools throughout Singapore.
- AgSphere: TelemetryWeb.com is launching AgSphere, a new platform that takes the complexity and pain out of connecting agricultural technology products to the web quickly and at low cost. Manufacturers of agricultural equipment can build web-connected solutions that increase margins, reduce risk, and improve efficiencies for farmers by harvesting information from the farm.

6.1.1 Two Pillars of the Web

The invention of HTML/HTTP/URL on top of TCP/IP-based Internet started the Internet revolution; however, it was not until the killer application—Netscape web browser surfaces—that the Internet revolution, symbolized by the World Wide Web, really took off. The Netscape web browser evolved from the earlier Mosaic web browser. It was co-authored by Marc Andreessen at the National Center for Supercomputing

Applications of the University of Illinois Urbana–Champaign beginning in late 1992 and released in 1993. Mosaic was also a client for earlier protocols such as file transfer protocol (FTP), network news transfer protocol (NNTP), and gopher, but HTTP with HTML/URL ruled at the end.

On the other front, the application server became the foundation that helped build widely spreading web-based applications. An application server is a software framework or middleware that provides an environment in which applications can run, no matter what the applications are or what they do. An application server acts as a set of components accessible to the software developer through an API defined by the middleware itself. For web applications, these components are usually performed in the same machine where the web server is running, and their main job is to support the construction of dynamic web pages. However, present-day application servers target much more than just web page generation: they implement services like clustering, fail-over, and load balancing, so developers can focus on implementing the business logic.

The application server is based on the three-tiered (Figure 6.2) or multitiered software architecture. The multitier architecture is a client–server architecture in which the presentation, the

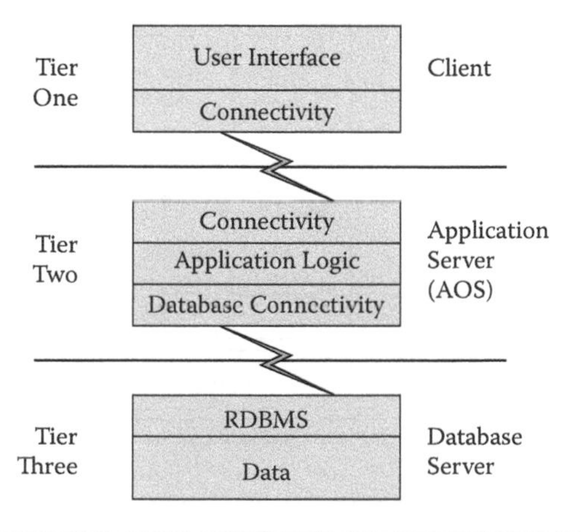

Figure 6.2 Three-tiered architecture.

application processing, and the data management are logically separate processes, which is important for distributed web applications. For example, a web application that uses middleware to service data requests between a user and a database employs multitier architecture. The most widespread use of multitier architecture is the three-tier architecture, which was first used by John Donovan for open-standards Distributed Computing Environment–based applications in Open Environment Corporation, a tools company he founded in the early 1990s.

The Java technologies developed rapidly in parallel with the web in each and every aspect. The Java EE standard-based application server architecture is shown Figure 6.3, which dominates the overall application server market as shown in the Gartner Quadrant [232].

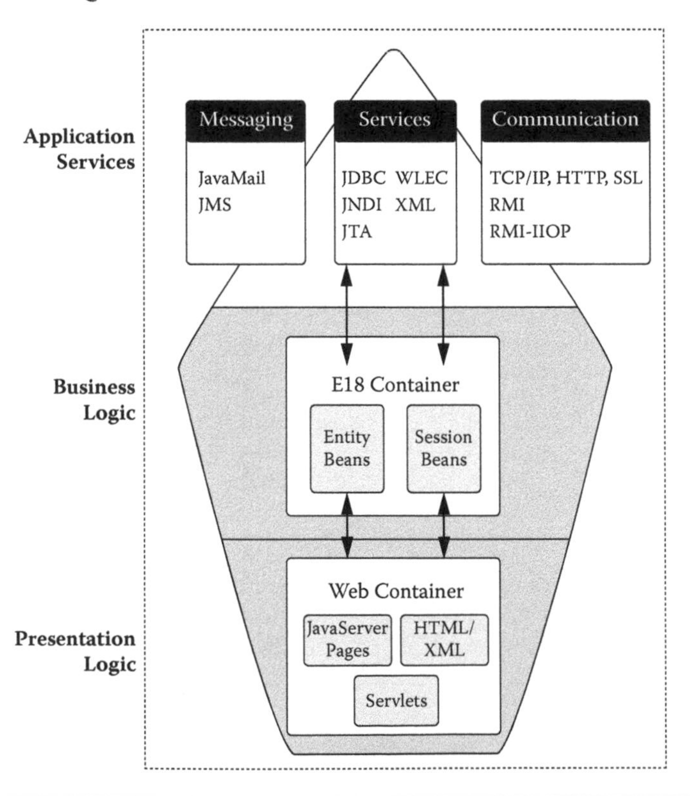

Figure 6.3 Java-based application servers.

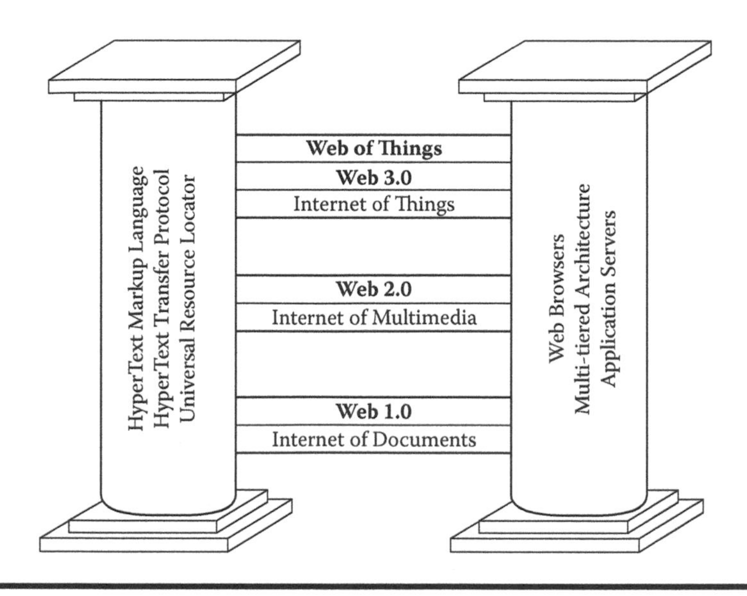

Figure 6.4 Two pillars of the web.

As the two pillars for web applications and the Internet revolution, the protocols (i.e., HTML now in its fifth version, HTML5)/HTTP/URL and the software (i.e., the web browsers and the standardized three-tiered application servers) will continue to be the two pillars of and play an important role in building WoT applications as depicted in Figure 6.4.

However, just as the web applications get more and more sophisticated, the HTML standard evolves, and a large number of standards and substandards and APIs (application programming interfaces) have been created, for example, for JavaEE and JavaME platforms. (These platforms are very relevant to IoT or machine-to-machine [M2M] applications, http://www.m2marchitect.com/what-is-m2m--2.html. There are Java Virtual Machines for all kinds of devices: JVM, CVM, KVM, CardVM, etc., as shown in Figure 6.5.) There is a need to update or augment those standards to fit the specific requirements of WoT/IoT applications, just like the wireless community has done for machine-type communication (MTC).

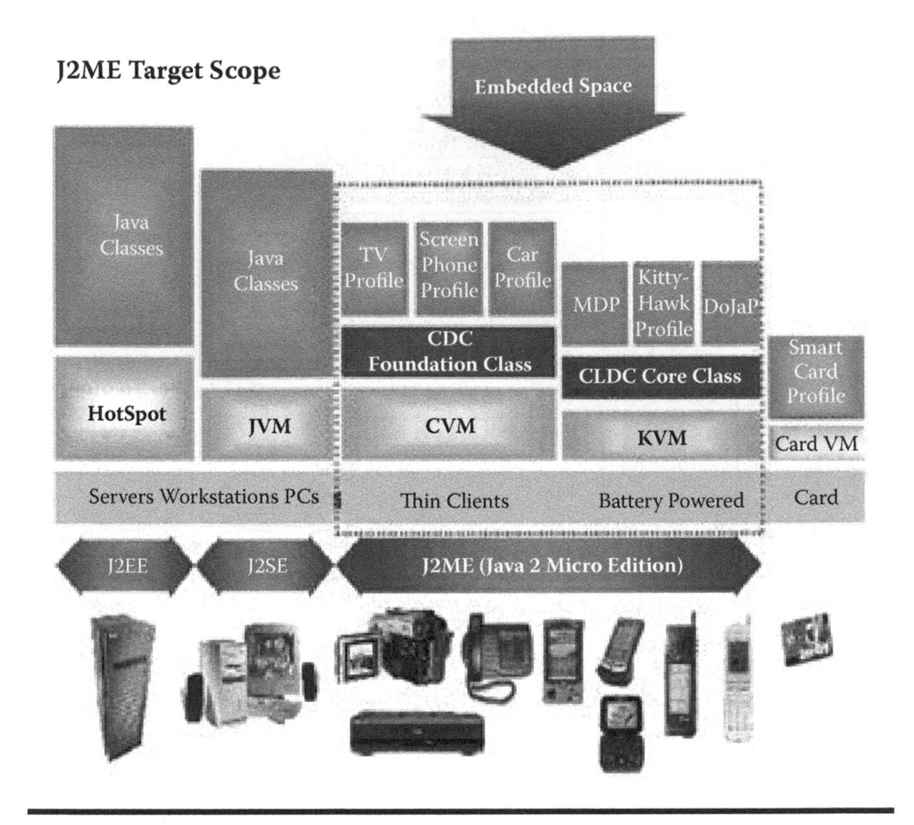

Figure 6.5 Java variants for devices.

A communications protocol is a language of digital message formats and rules for exchanging those messages in or between computing systems and/or in telecommunications. Protocols may include signaling, authentication, and error detection and correction capabilities. A protocol definition defines the syntax, semantics, and synchronization of communication. The specified behavior is typically independent of how it is to be implemented. A protocol can therefore be implemented as hardware or software or both.

Example protocols include data formats like HTML, ebXML (electronic business or e-business extensible markup language), and communication rules (or loosely called protocols) such as SOAP (simple object access protocol) and REST (representational state transfer). We will talk about horizontal and data format (like HTML) and protocol standard efforts for WoT/IoT applications in greater detail (and propose a unified data representation approach) in this chapter, and WoT/IoT specific middleware and multitiered architecture standard efforts in the next chapter.

6.2 IoT Protocol Standardization Efforts

We have touched on the issues of IoT standardization sporadically in the previous chapters of the book. Now we are going to give a summarized description of the four pillars as well as the generic IoT standardization efforts focusing on data representations and APIs (i.e., protocols). The standards on platform architecture and middleware framework will be discussed in the next chapter. However, because in most cases, the data representation and APIs are intertwined with architecture and framework, it is hard to separate; so there may be some overlaps.

Some of the IoT projects such as the Internet of Things Strategic Research Roadmap by CERP-IoT [8] are still at the grand concept level with limited materialized results. The IoT-A (Internet of Things architecture [113]) is one of the few efforts targeting a holistic architecture for all IoT sectors. This consortium consists of 17 European organizations from nine countries. They summarized the current status of IoT standardization as follows:

- Fragmented architectures, no coherent unifying concepts, solutions exist only for application silos.
- No holistic approach to implement the IoT has yet been proposed.
- Many island solutions do exist (RFID, sensor nets, etc.).
- Little cross-sector reuse of technology and exchange of knowledge.

The author had the same observation (also one of the first who introduced the Intranet/Extranet of Things concept independently [74]) before 2010 based on the four-pillar classification of IoT. Even though the IoT-A consortium doesn't categorize the IoT as four pillars, they do believe solutions for radio-frequency identification (RFID), sensor nets, and so forth are island solutions. In fact, IoT-A doesn't have a systematic, clean-cut, and comprehensive classification of IoT sectors as the foundation. Their "holistic" view of IoT is based on the following scenarios, which is actually not complete and holistic currently.

The key objectives of the IoT-A consortium [103] are as follows:

- Create the architectural foundations of an interoperable Internet of Things as a key dimension of the larger future Internet
- Architectural reference model together with an initial set of key building blocks:
  - Not reinventing the wheel but federating already existing technologies
  - Demonstrating the applicability in a set of use cases
  - Removing the barriers of deployment and wide-scale acceptance of the IoT by establishing a strongly involved stakeholder group
- Federating heterogeneous IoT technologies into an interoperable IoT fabric

A WP (work package) framework of ongoing works has been proposed [103]. Also, the ITU-T has a few study groups (SGs 2, 3, 5, 9, 11, 12, 13, 15, 16, and 17, http://www.itu.int/en/ITU-T/techwatch/Pages/internetofthings.aspx) doing IoT-related works (Figure 6.6).

IPSO (Internet Protocol for Smart Objects, http://www.ipso-alliance.org/) Alliance, formed in 2008, is another effort aiming to form an open group of companies to market and educate about how to use IP for IoT smart objects based on an all-IP holistic approach [81] (Figure 6.7).

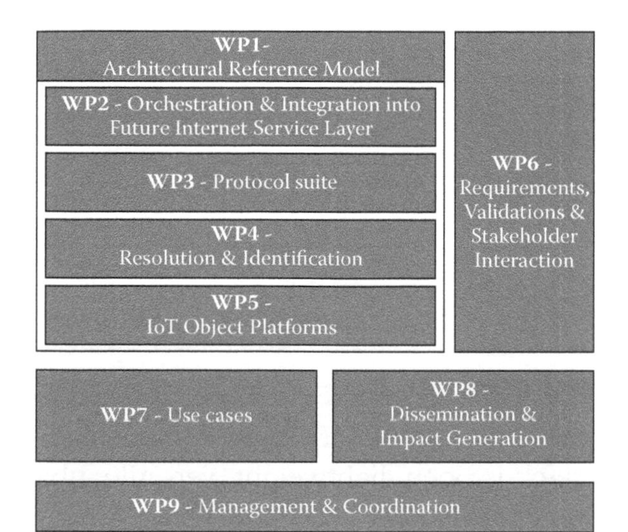

	ITU-T Study Group	Study Group Name	Activities Related to IoT
Current Standards Activities	SG 2	Operational aspects of service provision and telecommunication management	Numbering, naming and addressing
	SG 3	Tariff and accounting principles including related telecommunication economic and policy issues	
	SG 5	Environment and climate change	
	SG 9	Television and sound transmission and integrated broadband cable networks	
	SG 11	Signalling requirements, protocols and test specifications	Testing architecture for tag-based identification systems and functions
	SG 12	Performance,QoS and QoE	
	SG 13	Future networks including mobile and NGN	NGN requirements and architecture for applications and services using tag-based ID
	SG 15	Optical transport networks and access network infrastructures	
	SG 16	Multimedia coding, systems and applications	Requirements and architecture for multimedia information access triggered by tag-based ID
	SG 17	Security	Security and privacy of tag-based applications
Pre-standards	Focus Groups	Smart Grid	Smart metering, M2M
		Cloud Computing	Cloud network requirements, e.g., for IoT
		Future Networks	Describe future networks underlying the IoT
		Car Communication	

Figure 6.6 Working groups of IoT standards.

The emerging application space for smart objects requires scalable and interoperable communication mechanisms that support future innovation as the application space grows. IP has proven itself a long-lived, stable, and highly scalable communication technology that supports a wide range of

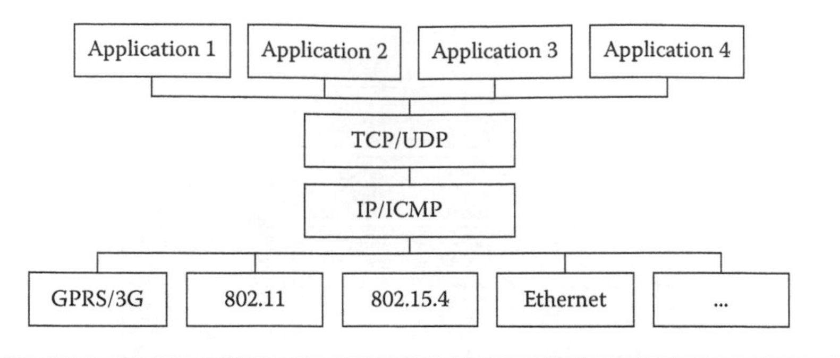

Figure 6.7 All-IP networks.

applications, devices, and underlying communication technologies. The IP stack is open, lightweight, versatile, ubiquitous, scalable, manageable, stable, and end-to-end. It can run on tiny, battery-operated embedded devices. IP therefore has all the qualities to make the Internet of Things a reality, connecting billions of communicating devices. A smart object is defined by IPSO as

- An intelligent (RFID) tag
- A sensor: device that measures a physical quantity and converts it to an analog or digital signal, such as power consumption and quality, vibration of an engine, pollution, motion detection, temperature
- An actuator: device that controls a set of equipment, such as controls and/or modulates the flow of a gas or liquid, controls electricity distribution, performs a mechanical operation
- An embedded device: a purpose-built connected device that performs a specific function, such as a factory robotic arm, vending machine, smart grid analyzer
- Any combination of the above features to form a more complex entity

The IPSO Alliance works closely with Internet Engineering Task Force (IETF), the Institute of Electrical and Electronics

Engineers (IEEE), the European Telecommunication Standard Institute (ETSI), the International Society of Automation (ISA), and others, and relies on the standards developed by them. IPv4, IPv6, and 6LoWPAN were all developed by engineers within IETF, and the role of the alliance is to ensure how they are used, deployed and provided to all potential users.

The Mobile IP protocol is a related IETF-proposed standard that provides a network layer solution to node mobility across IPv4 (Mobile IPv4) and IPv6 (Mobile IPv6) networks. Mobile IP allows a node to change its point of attachment to the Internet without having to change its IP address.

Another solution to the problem is network mobility (NEMO). NEMO is an extension of Mobile IP that enables an entire network to change its attachment point to the Internet. NEMO works by moving the mobility functionality from Mobile IP mobile nodes to a moving network's router. The router is able to change its attachment point to the Internet in a manner that is transparent to attached nodes.

SHIM6 [114], a serverless Mobile IPv6 protocol, allows two communicating nodes to overcome connection loss problems that may arise if one node changes its IP address (locator) during an established communication.

Sensinode [115], as an example, provides embedded networking software and hardware products based on IP-based 6LoWPAN technology for demanding enterprise applications. NanoStack™ 2.0 is an advanced 6LoWPAN protocol stack software product for 2.4 GHz radios. The NanoRouter™ 2.0 platform includes software and hardware solutions for 6LoWPAN-Internet routing infrastructure.

Also, since its creation in 2003, ETSI TISPAN (Telecommunications and Internet converged Services and Protocols for Advanced Networking) has been the key standardization body in creating the next-generation networks (NGN) specifications, which is a synonym of IoT.

6.2.1 M2M and WSN Protocols

Most M2M applications are developed today in a highly customized fashion, and vertical-specific industry bodies are busy crafting standards for markets ranging from the auto industry to the smart grid. A broad horizontal standard is a key requirement for the M2M industry to move from its current state of applications existing in isolated silos based on vertical market or underlying technology to a truly interconnected Internet of Things. Such a horizontal standard is expected to be the major impetus to growth in the future.

Efforts to develop broad, horizontal standards for the M2M market are gaining momentum [49,105]. The most important activity is occurring within the context of the International Telecommunication Union's (ITU) and ETSI's (M2M Technical Committee) Global Standards Collaboration (GSC), which has established the M2M Standardization Task Force (MSTF, created during the GSC-15 meeting in Beijing, China, in September 2010) to coordinate the efforts of individual standards development organizations (SDOs), including China Communications Standards Association, Telecommunications Industry Association TR-50 Smart Device, etc.

The end result of these efforts is to define a conceptual framework for M2M applications that is vertical industry and communication technology agnostic, and to specify a service layer that will enable application developers to create applications that operate transparently across different vertical domains and communication technologies without the developers having to write their own complex custom service layer [105]. The high-level M2M architecture from MSTF does include fixed and other noncellular wireless networks, which means it's a generic, holistic IoT architecture even though it is called M2M architecture (M2M and IoT sometimes are used interchangeably in the United States and in the telco-related sectors). Despite all of the positives, it seems the voices from the SCADA (supervisory control and data acquisition) and RFID communities are relatively weak; efforts to incorporate existing SCADA standards such as OPC, ISA-95, and RFID EPCIS, ONS, and others are not seen yet. It remains to be seen whether all of the stakeholders from the four pillars of IoT will be equally included in the loop.

This is a more comprehensive approach than the 3GPP's MTC effort described in the previous chapter. Considering 3GPP is only one of the SDOs in the MSTF, this makes sense and good results are much anticipated from MSTF. Some vertical applications on top of the unified horizontal M2M architecture are already under way [105]. Companies such as Telenor Objects, Numerex, and others are building MSTF standards compliant products [104] already.

Other M2M standards activities include the following:

- Data transport protocol standards: M2MXML, JavaScript Object Notation (JSON) (originally not for IoT applications, used by the Mango open source M2M project), BiTXML [117], WMMP (shown in Figure 6.8), MDMP, open Building Information Exchange (oBIX), EEML, open M2M Information exchange (oMIX)
- Extend OMA DM to support M2M devices protocol management objects
- M2M device management, standardize M2M gateway
- M2M security and fraud detection
- Network API's M2M service capabilities
- Charging standards
- MULTI IMSI, M2M sevices that do not have MSISDN
- IP addressing issues for devices IPV6
- Remote diagnostics and monitoring, remote provisioning and discovery
- Remote management of devices behind a gateway or firewall
- Open REST-based API for M2M applications

One of the benefits of using sensor data is that the data typically can be repurposed many times, thereby reducing cost

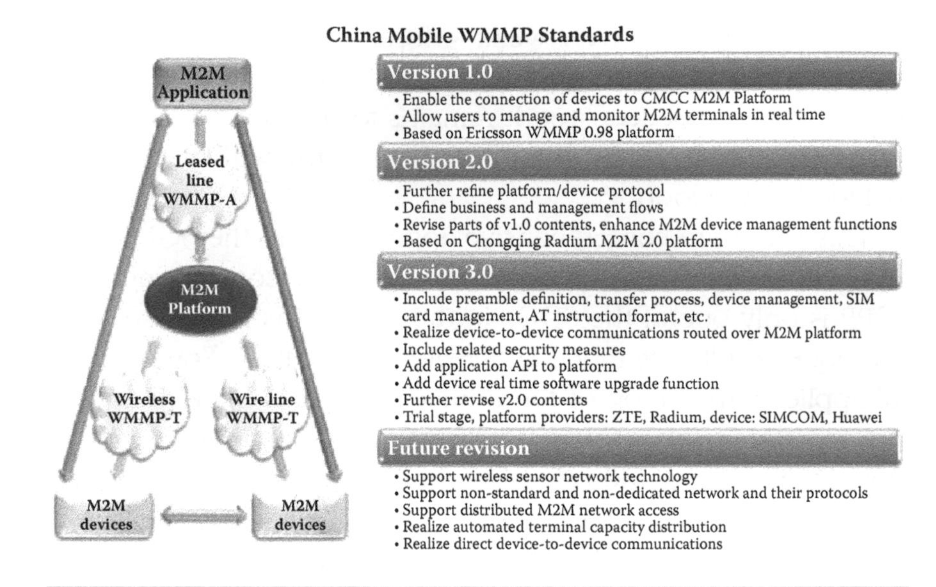

Figure 6.8 China Mobile's WMMP standard.

and maximizing benefit. For example, weather observations (temperature, wind speed and direction, humidity, and so on) can be used in climate modeling, weather forecasting, plume modeling, insurance risk analysis, ski area location decisions, and dozens of other applications. However, the ability to access and use the same sensors in multiple application domains, to share sensor data, and to maximize the full value of sensor networks and data is severely hindered by a lack of interoperability. Hundreds of sensor manufacturers build sensors for specific purposes, often using their own "language" or encodings, different metadata, and so forth. Standard data representation (together with WSN middleware) is the key to materialize data integration and increase interoperability.

There are a number of standardization bodies in the field of WSNs. The IEEE focuses on the physical and MAC layers; the IETF works on layers 3 and above. IEEE 1451 is a set of smart transducer interface standards developed by the IEEE Instrumentation and Measurement Society's Sensor Technology Technical Committee that describe a set of open, common, network-independent communication interfaces for connecting transducers (sensors or actuators) to microprocessors, instrumentation systems, and control/field networks. One of the key elements of these standards is the definition of transducer electronic data sheets (TEDS) for each transducer. The TEDS is a memory device attached to the transducer, which stores transducer identification, calibration, correction data, and manufacturer-related information.

The IEEE 1451 family of standards includes the following:

- 1451.0-2007 Common Functions, Communication Protocols, and TEDS Formats
- 1451.1-1999 Network Capable Application Processor Information Model
- 1451.2-1997 Transducer to Microprocessor Communication Protocols & TEDS Formats
- 1451.3-2003 Digital Communication & TEDS Formats for Distributed Multi-drop Systems
- 1451.4-2004 Mixed-mode Communication Protocols & TEDS Formats
- 1451.5-2007 Wireless Communication Protocols & TEDS Formats
- 1451.7-2010 Transducers to Radio Frequency Identification (RFID) Systems Communication Protocols and TEDS Formats

The goal of the IEEE 1451 family of standards is to allow the access of transducer data through a common set of interfaces whether the transducers are connected to systems or networks via a wired or wireless means. IEEE p1451.3 is XML based and allows the manufacturer to change the contents.

Cross-network (e.g., between Bluetooth and ZigBee) standards are not as proliferate in the WSN community compared to other computing systems, which make most WSN systems incapable of direct communication with each other. The contents on WSN described in the previous chapters are more devices or network focused. OGC (Open Geospatial Consortium) and W3C has been doing research and standardization work following a data-focused approach [233].

The Semantic Sensor Web (SSW) [105] is an approach to annotating sensor data with spatial, temporal, and thematic semantic metadata based on OGC SWE (Sensor Web Enablement). The following data-encoding specifications have been produced by OGC SWE Working Group (in addition to the web service specifications that will be described in Chapter 7):

- SWE Common—common data models and schema
- SensorML—models and schema for sensor systems and processes surrounding measurements
- Observations & Measurements (O&M)—models and schema for packaging observation values
- Transducer Markup Language (TML)—models and schema for multiplexed data from sensor systems

The European Union SENSEI [109] project creates an open, business driven architecture that fundamentally addresses the scalability problems for a large number of globally distributed wireless sensor and actuator networks (WSAN) devices. It provides necessary network and information management services to enable reliable and accurate context information retrieval and interaction with the physical environment. By adding mechanisms for accounting, security, privacy, and trust, it enables an open and secure market space for context awareness and real-world interaction. An ambient ERP system supported the SENSEI.

Tangible results of the SENSEI project are as follows:

- A highly scalable architectural framework with corresponding protocol solutions that enable easy plug-and-play integration of a large number of globally distributed WSAN into a global system, providing support for network and information management, security, privacy and trust, and accounting

- An open service interface and corresponding semantic specification to unify the access to context information and actuation services offered by the system for services and applications
- Efficient WSAN island solutions consisting of a set of cross-optimized and energy-aware protocol stacks including an ultra-low-power multi-mode transceiver
- Pan European test platform, enabling large-scale experimental evaluation of the SENSEI results and execution of field trials, providing a tool for long-term evaluation of WSAN integration into the NGN

ISO/IEC JTC1 WG7 (Working Group on Sensor Networks), established in 2009, preceded by JTC 1 SGSN SC6, created the ISO/IEC 29182 Reference Architecture for sensor networks application and services focusing on telecommunication and information exchange between systems. The architecture is defined through the following set of documents:

- ISO/IEC 29182 Part 1: General overview and requirements
- ISO/IEC 29182 Part 2: Vocabulary/terminology
- ISO/IEC 29182 Part 3: Reference architecture views
- ISO/IEC 29182 Part 4: Entity models
- ISO/IEC 29182 Part 5: Interface definitions
- ISO/IEC 29182 Part 6: Application profiles
- ISO/IEC 29182 Part 7: Interoperability guidelines

6.2.2 SCADA and RFID Protocols

As described before, we use the SCADA term as one of the IoT pillars to represent the whole industrial automation arena in this book. Industrial automation has a variety of vertical markets and there are also many types of SCADAs.

IEEE created a standard specification, called Std C37.1™, for SCADA and automation systems [116] in 2007, targeting mostly

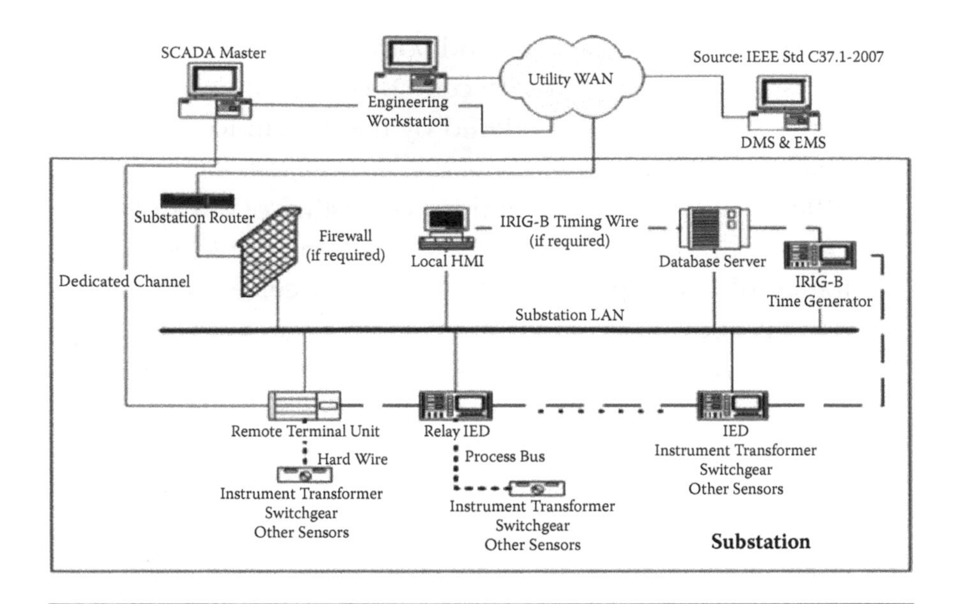

Figure 6.9 IEEE Std. C37.1 SCADA architecture.

power SCADA applications (Figure 6.9). It's recognized in the specification that in recent years, network-based industrial automation has greatly evolved with the use of intelligent electronic devices (IEDs), or IoT devices in our terms, in substations and power stations. The processing is now distributed, and functions that used to be done at the control center can now be done by the IED, that is, M2M between devices. Despite the fact that many functions can be moved to the IED, utilities still need a master station, the IoT platform, for the operation of the power system. Due to the restructuring of the electric industry, traditional vertically integrated electric utilities are replaced by many entities such as GENCO (Generation Company), TRANSCO (Transmission Company), DISCO (Distribution Company), ISO (independent system operator), RTO (regional transmission organization), and so forth. To fulfill their role, each of these entities needs a control center, that is, a substation, to receive and process data and take appropriate control actions.

This specification addressed all levels of SCADA systems and covered the technologies used and, most importantly, the architecture of how those technologies interact and work together. However, no XML data formats and componentized architecture details are specified, which is perhaps why SCADA has long been regarded as a traditional control system market. People working in that area are often not aware of Internet-based IT innovations and cannot relate their work to a new concept such as IoT.

Wireless sensor systems have the potential to help industry use energy and materials more efficiently, lower production costs, and increase productivity. Although wireless technology has taken a major leap forward with the boom in wireless personal communications, applications for industrial field device systems must meet distinctly different challenges. That's where the ISA100, Wireless Systems for Industrial Automation, comes in. The ISA100 was developed by the standards committee of the Industrial Society for Automation, which was formed in 2005 to establish standards and related information that will define procedures for implementing wireless systems in the automation and control environment with a focus on the field level. The committee is made up of more than 400 automation professionals from nearly 250 companies around the world, lending their expertise from a variety of industrial backgrounds.

The ISA100 family of standards is designed with coexistence in mind, bringing peace of mind for the end user. We know that customers have other wireless solutions installed today and have the need for any future system to coexist with these installed systems. Therefore, the standards will feature technology to ensure the best performance possible in the presence of other wireless networks. For example, the ISA100 has created a new subcommittee to address options for convergence of the ISA100.11a and WirelessHART standards. This initiative is a key step in the mission of the ISA100 committee to develop a family of universal industrial wireless standards designed to satisfy the needs of end users across a variety of applications.

OPC, which stands for Object Linking and Embedding (OLE) for Process Control, is the original name for a standard

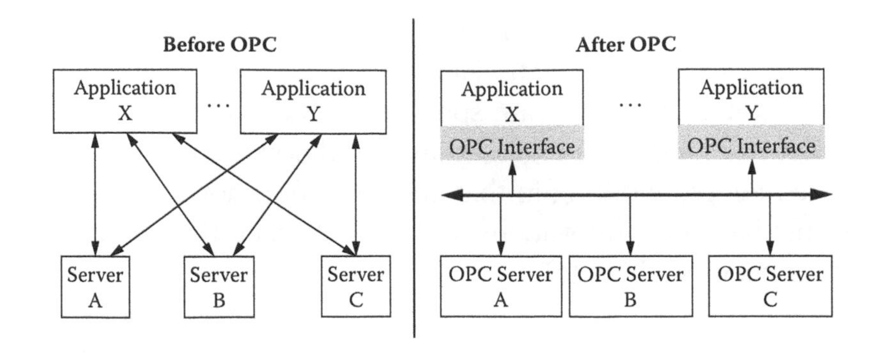

Figure 6.10 OPC standard for device connection.

specification developed in 1996 by an industrial automation industry task force. The standard specifies the communication of real-time plant data between control devices from different manufacturers (Figure 6.10). OPC is managed by the OPC Foundation [120] with more than 220 members worldwide including major firms in industrial automation, instruments manufacturers, building automation, and others.

OPC originated from the DDE (dynamic data exchange) technologies based on DOS for PCs. The introduction of Windows 3.0 in 1990 made Windows an inexpensive, mainstream computing platform, providing the ability for a PC to run multiple applications simultaneously and a standard mechanism for those applications to exchange data at runtime. Wonderware's InTouch™ SCADA software had the greatest impact for the transition from DDE to OPC. It introduced a means of networking DDE traffic (NetDDE™, which was later taken up by Microsoft) and also greatly increased the effective bandwidth of DDE by packing multiple data items into each packet or message. OLE (based on COM, common object model) and OCX (now ActiveX based on .NET) were launched in 1992. A number of SCADA vendors saw the chance to standardize the interface between the SCADA core and the device drivers that were actually responsible for acquiring the data, and the first-draft version of the OPC specification was released in December 1995 by the OPC Foundation sponsored by Microsoft.

OPC was designed to provide a common bridge for Windows-based software applications and process control hardware. Standards define consistent methods of accessing field data from plant floor devices. This method remains the same regardless of the type and source of data. An OPC server for one hardware device provides the same methods for an OPC client to access its data as each and every other OPC server for that same or another hardware device. The aim was to reduce the amount of duplicated effort required from hardware manufacturers and their software partners, and from the SCADA and other HMI producers, in order to interface the two. When a hardware manufacturer had developed their OPC server for the new hardware device, their work was done to allow anyone to access their device; and when the SCADA producer had developed their OPC client, their work was done to allow access to any hardware, existing or yet to be created, with an OPC-compliant server.

OPC has achieved great success in many application areas, most of them closely related to or part of IoT applications. However, OPC's success story is accompanied by some caveats. For example, standard OPC DA (data access) is based on Microsoft's COM and DCOM technology and is consequently restricted to the Windows operating system. In addition, DCOM communication is easily blocked by firewalls that prevent OPC clients from accessing data over a wide-area network and the World Wide Web. New approaches, such as XML-DA and United Architecture (UA) [234], have been developed to make OPC technology available on other platforms or accessible by other systems.

The RFID protocols and data formats are relatively well defined, mostly by EPCglobal, and unified compared with protocols and formats of the other three pillars of IoT. The RFID protocols (such as PML, Object Naming Service [ONS],

Edgeware, EPC Information Service [EPCIS], Application Level Event [ALE], etc.) have been described in the previous chapters, so we will talk only about protocols for the related contactless smart cards here.

The smart cards with contactless interfaces (RFID is a subset) are becoming increasingly popular for payment and ticketing applications such as mass transit and stadiums. Visa and MasterCard have agreed to an easy-to-implement version deployed in the United States. Smart cards are also being introduced in personal identification and entitlement schemes at regional, national, and international levels. Citizen cards, drivers' licenses, and patient card schemes are becoming more prevalent. Some examples of widely used contactless smart cards are Taiwan's EasyCard, Hong Kong's Octopus card, Shanghai's Public Transportation Card, and Beijing's Municipal Administration and Communications Card.

The standard for contactless smart card communications is ISO/IEC 14443. It defines two types of contactless cards (A and B) and allows for communications at distances up to 10 cm. An alternative standard for contactless smart cards is ISO/IEC 15693, which allows communications at distances up to 50 cm (Figure 6.11).

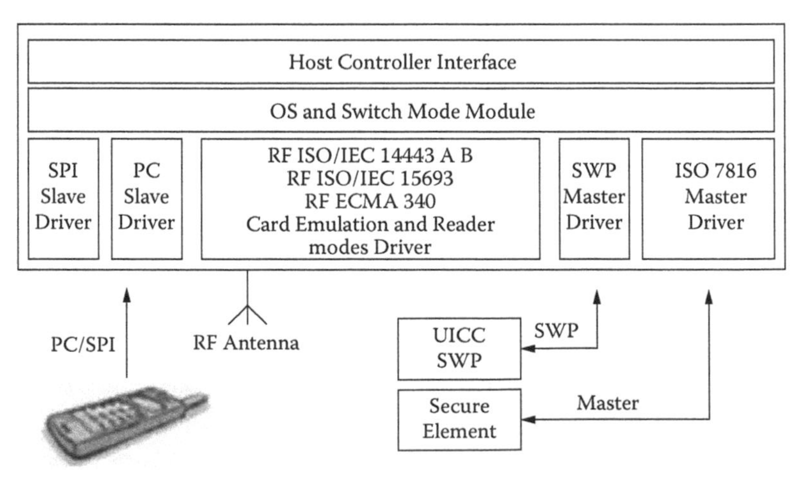

Figure 6.11 ISO/IEC 14443/15693 smart card standards.

6.2.3 Issues with IoT Standardization

Apart from the standardization efforts that can be categorized as one of four pillars, there are also standardization efforts from major vertical IoT applications such as smart grid and telematics. For example, we have described the NGTP (Next Generation Telematics Protocol) standard for telematics in Chapter 2. The participating SDOs of GridWise (Smart Grid) include almost all of the SDOs in the information and communications technology (ICT) industry.

It should be noted that not everything about standardization is positive. Standardization is like a double-edged sword: critical to market development, but it may threaten innovation and inhibit change when standards are accepted by the market. Standardization and innovation are like yin and yang, and they could be contradictory to each other in some cases, even though this observation is debatable.

We have also noted in the previous chapter that among the four pillar segments of IoT, for example, in ETSI/3GPP's M2M/MTC and EPCglobal's RFID standardization efforts, different consortia, forums, and alliances have been doing standardization in their own limited scope covering mostly the area they are familiar with. For example, 3GPP covers only cellular wireless networks and EPCglobal's middleware covers only RFID events. Even within the same segment, there are more than one consortium or forum doing standardization without enough communication with each other, and some are even competing with each other.

Some people believe that the IoT concept is well established; however, some gray zones remain in the definition, especially on which technologies should be included, such as the four pillars described in this book, in order not to pose a limit too strict to the system.

Even though some of the IoT standard organizations have cooperation and interaction, as shown in Figure 5 of Jacobs et al. [102], it is limited and not open enough. The following two issues for the IoT standardization in particular and the ICT standardization in general may never have answers:

- ICT standardization is a highly decentralized activity. How can the individual activities of the network of extremely heterogeneous standards-setting bodies be coordinated?
- It will become essential to allow all interested stakeholders to participate in the standardization process toward the IoT and to voice their respective requirements and concerns. How can this be achieved?

The only, or at least better, possible solution to address these chaotic situations is to try to standardize the omnipresent middleware and the XML-based data representation from across-industry organizations such as World Wide Web Consortium (W3C), Organization for the Advancement of Structured Information Standards (OASIS), and others.

OASIS and W3C are web-oriented standard organizations. Their expertise makes them capable of doing high-level, segment-independent WoT standardization. They are now actually participants of ETSI/3GPP and other efforts, but they are currently more like observers instead of active participants. Most other IoT SDOs are more qualified to do IoT (communication layers) standardization instead of WoT standardization since they often lack a high-level view and experiences of the system across the globe and across industries.

6.3 Unified Data Standards: A Challenging Task

We have talked about the two pillars of the Internet in this and previous chapters and pointed out that the HTML/HTTP combination of data format and exchange protocol is the foundation pillar of the World Wide Web as depicted in Figure 6.12 [74].

We have also listed and described a great number of data standards and protocols proposed for the four pillar domains

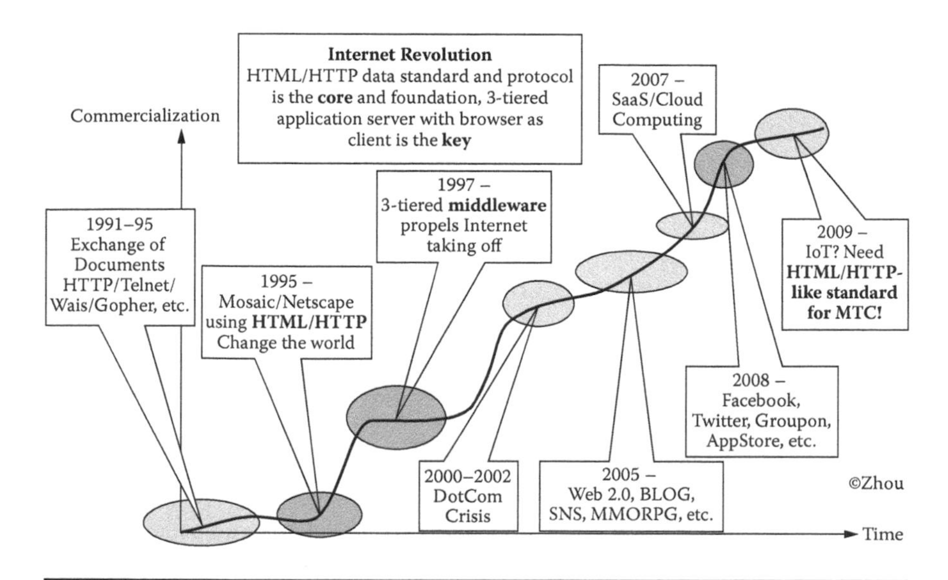

Figure 6.12 Evolution of the web.

of the IoT in the previous sections of this chapter and previous chapters. Many issues still impede the development of IoT and especially the WoT vision.

Many standardization efforts have been trying to define a unified data representation and protocol for IoT. This is the right direction even though some of the approaches are limited by their scope of application domains and technologies used as described and discussed before.

Before the Internet of Things, the Internet was actually an Internet of documents or of multimedia documents. The two pillars of the Internet including HTML/HTTP turned the Internet into the World Wide Web. By the same token, we need to turn the Internet of Things into the Web of Things to make sense of everything. What will it take to make this to happen?

- Do we need a new HTML/HTTP-like standard for MTC and WoT? If there is no need to reinvent the wheel, what extensions do we need to build on top of HTML/HTTP or HTML5?

- The browser is intended for humans, so do we need a new browser for machines to make sense of the ocean of machine-generated data? If not, what extensions do we need to make to the existing browsers?
- Today, most new protocols are built on top of XML. For OS there must be XML-based data format standards or a metadata standard to represent the machine-generated data (MGD). Is it possible to define such a metadata standard that covers everything?

There are many different levels of protocols, but the ones that most directly relate to business and social issues are the ones closest to the top, the so-called application protocols such as HTML/HTTP for the web. The web has always been a visual medium, but a restricted one at best. Until recently, HTML developers were limited to CSS and JavaScript in order to produce animations or visual effects for their websites, or they would have to rely on a plug-in like Flash. With the addition of technologies like the canvas element, Web GL, and scalable vector graphics (SVG) images in HTML5, this is no longer the case. In fact, many new features deal with graphics on the web with HTML5: 2-D Canvas, WebGL, SVG, 3-D CSS transforms, and Synchronized Multimedia Integration Language (SMIL).

Developers are taking advantage of these features: a flood of HTML graphics demos have been showing up on the web, ranging from implementations of old two-dimensional graphics algorithms, to brand-new techniques created specifically for the modern web. Using graphics to display real-world behavior of things is a very important feature of IoT systems; for example, sophisticated graphic display of device behavior and process is a must in most SCADA-based industrial automation systems. The use of SVG technologies can reduce the footprint of a graphic by up to 90 percent. On the left of Figure 6.13 is an oil and gas industrial automation application using SVG built by the author's team in 2008 before HTML5 was announced. A tool or IDE (integrated development

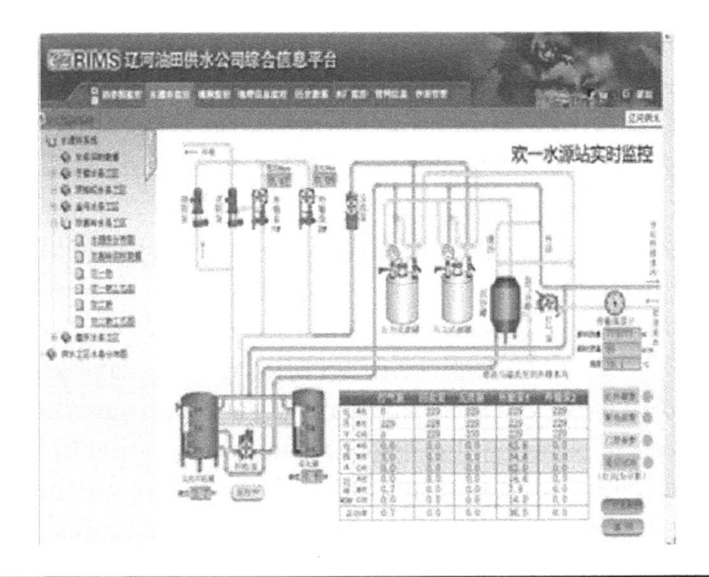

Figure 6.13 SVG graphics of ^{ez}M2M.

environment) is normally required as a companion product for SCADA to create and configure graphics in large volume efficiently and productively based on a large graphic (parts) library.

Mango is an open-source software system for M2M applications. It enables users to access and control electronic sensors, devices, and machines over multiple protocols simultaneously. It relies heavily on JavaScript to render its graphical pages. While rendering, massive amounts of data are being transferred between the Mango server and the browser. Furthermore, because of the continuous polling for new data, it can easily hog the central processing unit of the computer displaying said data. SVG, supported by HTML5 as well as major browsers such as Internet Explorer 9, Safari (Apple doesn't support Adobe Flash), and others, with embedded scripting capabilities can be a very useful technology; however, enhancement to HTML/HTTP and the browsers is still required for MTC support. Human-oriented browsers may also have to be enhanced for processing massive MGDs similar to the mobile browser on audio devices. Content management is a big market sector of the Internet and web; future IoT contents may also require similar technologies for sensor content management.

The Resource Description Framework (RDF) is a family of W3C specifications originally designed as a metadata model. It has come to be used as a general method for conceptual description or modeling of information that is implemented in web resources, using a variety of syntax formats. It could be investigated and used as a metadata model for WoT applications.

An RDF browser is a piece of technology that enables you to browse RDF data sources by way of data link traversal. The key difference between this approach and traditional browsing is that data links are typed (they possess inherent meaning and context just like IoT data), whereas traditional links are not typed. There are a number of RDF browsers including Tabulator, DISCO (Hyperdata Browser), and OpenLink RDF Browser.

SOAP and RESTful protocol frameworks are extensions on top of HTTP for web services. They are more than protocols or data formats but rather the so-called protocol frameworks. SOAP and REST frameworks can be used to provide data exchange protocols for IoT applications, which will be discussed in the next chapter.

At the back-end server side or deep down in the ETSI/3GPP-defined M2M/IoT protocol stack, a unified IoT data format and protocol can borrow and leverage the standards proposed for e-commerce or e-business, especially B2B (business-to-business) standards. To clarify, EAI (enterprise application integration) is the integration of legacy software systems within an organization to allow the systems to have a more complete and consistent worldview. This is essentially an internal matter. B2B is about cross-organization integration, the creation of public interfaces to allow partners and customers to interact with internal systems in a programmatic fashion.

E-commerce comprises the B2B, business-to-consumer (B2C), and consumer-to-consumer (C2C) business models, which describe who the target buyer market the target seller market are. B2B application integration bridged the gap between legacy IT infrastructures and emerging B2B collaboration frameworks and allows the IT infrastructure to provide greater adaptability to the business of the enterprise and easier management of constantly evolving business processes. The same principle and technologies apply to legacy IoT infrastructure and emerging Internet- and web-based IoT collaboration frameworks also.

There are two important enabling technologies: electronic data interchange (EDI) and XML. EDI describes the rigorously standardized format of electronic documents. The EDI standards were designed to be independent of communication and software technologies. EDI can be transmitted using any methodology agreed to by the sender and recipient. This includes a variety of technologies, including modem (asynchronous and synchronous), FTP, e-mail, HTTP, AS1, AS2, and so forth. XML is a more recent invention for exchanging information between computer systems. XML is a markup language used to create smart data and documents for applications.

A newer standard such as ebXML incorporates as part of its design solution some borrowed ideas from both EDI and XML. It offers businesses the opportunity to build an interoperable e-commerce infrastructure. In a computer system, ebXML specifies the business rules for how two different systems talk to each other. Those systems need to be written using any application programming language (such as XML, Java, C, C++, or Visual Basic), executed in a specific middleware (like JavaEE or COM+; the author has worked in the BEA Weblogic Integration team developing Java software frameworks based on ebXML and RosettaNet protocols for e-commerce applications), and designed using a specific modeling language (UML). To model B2B business processes, an abstract computer-modeling language such as UML or the XML language–specific business process modeling language (BPML) is used. BPML is an XML-based meta-language for modeling, deploying, and managing business processes such as order management, customer care, demand planning, product development, and strategic outsourcing.

XML or ebXML coexists with the popular web formatting language HTML. HTML tells us how the data should look, but XML tells us what it means. XML enables complex linking (using XPointer and XLink) and allows users to define their own elements (using a document type definition [DTD] or schema). It also provides a style sheet for formatting documents (using XSL). The key issue of IoT applications is also about integration and interoperability, so the HTML/ebXML approaches still apply and new HTML-based, ebXML-like standards should be the solution for the Internet of Missed Things and the focus of IoT data representation standards for WoT applications.

There are a few specifications for the WoT data format mentioned before. The following is a longer list, which is summarized in *Smarter Earth* [74]:

- BITXML, data format defined by BITX Inc.
- CBRN, format for Chemical, Biological, Radiological and Nuclear data
- CAP, Common Alerting Protocol, of EXDL
- EDDL, Electronic Device Description Language
- EEML, Extended Environments Markup Language from Pachube
- EXDL, Emergency Data Exchange Language of OASIS
- FDT, Field Device Tool
- IRIG, Inter-Range Instrumentation Group
- MDMP, M2M protocol of China Telecom
- M2MXML, Machine-to-Machine XML
- NGTP, Next-Generation Telematics Protocol
- oBIX, open Building Information eXchange
- OMA SyncML, Open Mobile Alliance Synchronization Markup Language
- oMIX, open Machine Information eXchange, proposed by the author's team
- OPC, OLE for Process Control
- PML, Physical Markup Language of EPCglobal
- SensorML, Sensor XML of OGC

- TEDS/IEEE 1451, transducer electronic data sheets of IEEE
- TransducerML, Transducer Markup Language of OGC
- WMMP, Wireless Machine Management Protocol of China Mobile

Figure 6.14 shows the example schema of oBIX (open Building Information eXchange).

There are many ongoing and in some cases overlapping efforts to develop the CBRN standards within industrial, federal,

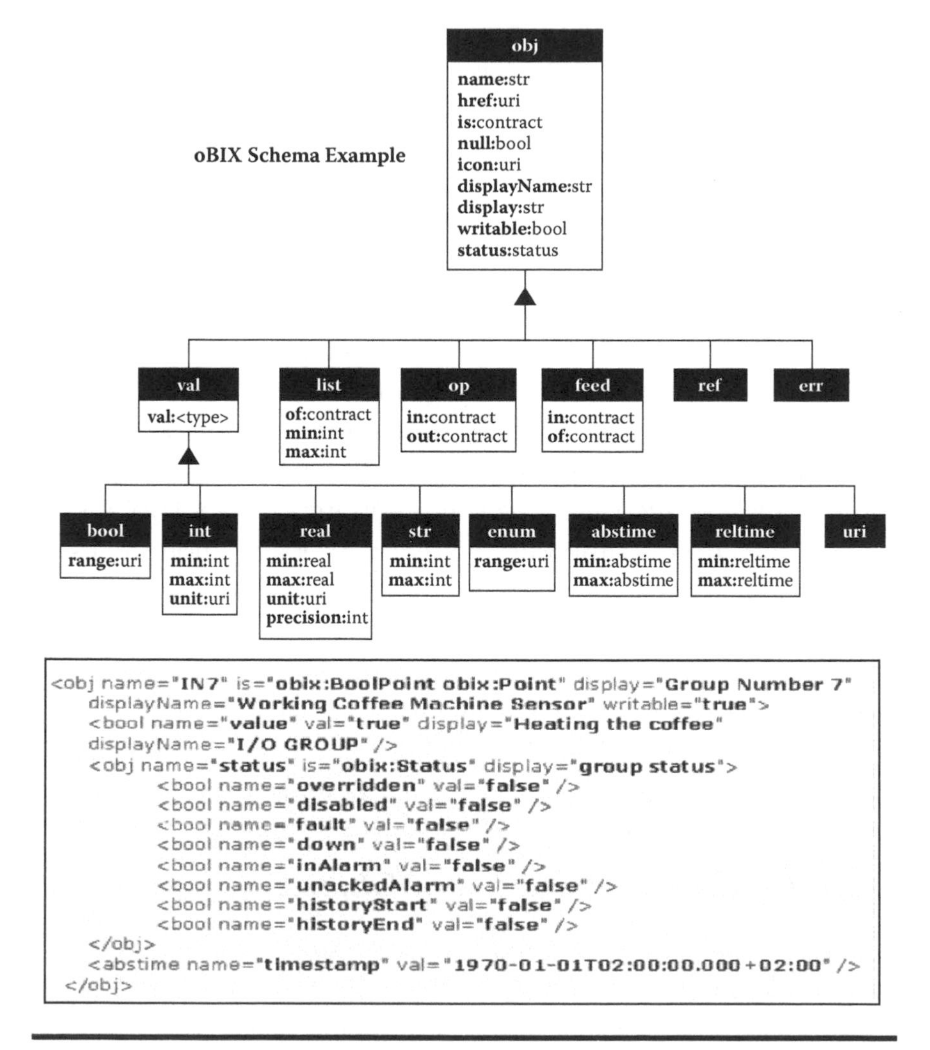

Figure 6.14 oBIX schema.

and international standards organizations. Oak Ridge National Laboratory, Tennessee (ORNL), where the author has worked, has invested a significant amount of research and development into implementing, testing, deconflicting, and harmonizing these efforts to establish an overarching set of working interoperability standards to connect CBRN sensors, detectors, and data to emergency response, homeland security, and defense applications (Table 6.1).

EDI for e-commerce is like OPC for WoT/SCADA—it's a legacy standard. A new, unified, open, cross-pillar, and usable standard like ebXML for WoT is needed, or efforts must be made to harmonize [123] the existing standards and to make them interoperable such as those in Figure 6.15 before a general sensor information model or a metadata XML schema can be established eventually.

Table 6.2 from *Smarter Earth* [74] summarizes the IoT data and protocol standardization efforts with the author's analysis, views, and suggestions about future developments.

As described before, there are many efforts to create a unified, cross-segment, overarching data representation standard for WoT. Due to domain knowledge differences, this is a great technological challenge or even mission impossible. Looking at the issues from a different angle, it is probably more realistic to create interoperability standards to integrate WoT systems between the four-pillar IoT systems. Even within a pillar segment, it's not an easy task to create a unified data standard. However, it's worth a try, especially at the early development stage of WoT before the IoT "information islands" are formed, as is the situation in many existing IT systems.

6.3.1 Unified Identification of Objects

One of the key issues of unified data format for IoT is the unique identification of objects. When the IoT application is within the intranet or extranet of an organization, which is the case most often currently, the identification is not an issue.

Table 6.1 CBRN Data Standard Efforts: Standards Activities for CBRN Sensors

Department of Defense	*DHS*	*Institute of Electrical and Electronics Engineers (IEEE)*	*OASIS*	*Open Geospatial Consortium (OGC)*
POC Activity				
JPEO-CBD	Standards Portfolio S&T Directorate	Sensor Interface Standards	Emergency Interoperability Consortium	Sensor Web Enablement
Prof. Tom Johnson, NPS	Dr. Bert Coursey, DHS S&T	Mr. Kang Lee, NIST	Ms. Elysa Jones, OASIS	Mr. Sam Bacharach, OGC
Standards				
JPM-IS Data CBRN	ANSI N42.32 ANSI N42.33	IEEE 1451.0 IEEE 1451.1	Common Alerting Protocol	Sensor Observation Services
Common Data Model	ANSI N42.34 ANSI N42.35 ANSI N42.38	IEEE 1451.2 IEEE 1451.3 IEEE 1451.4	Emergency Data	Sensor Planning Service Sensor Alerting Service
NATO NBC Standards (Allied Tactical Publication 45B)	ANSI N42.42 ASTM E54	IEEE 1451.5 IEEE 1451.6	Exchange Language	Geospatial Markup Language
STANAG 5523	AOAC International			Web Feature Services

Source: Courtesy of ORNL.

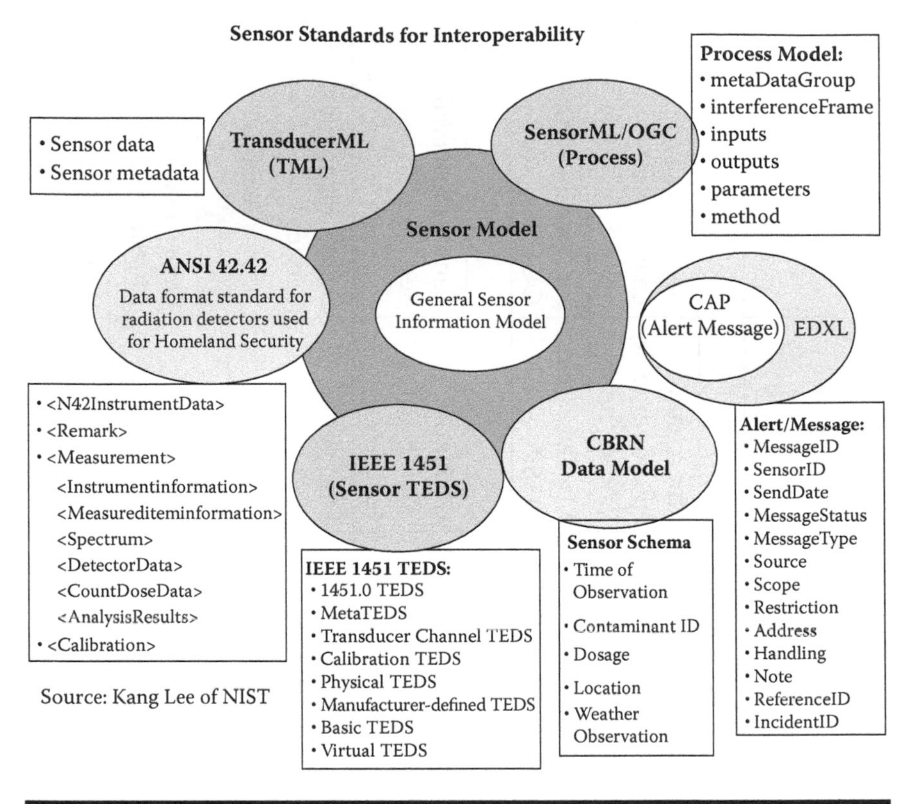

Figure 6.15 Unified Data Standard approaches.

However, when the WoT applications prevail in the future, globally unique identification of objects becomes a serious issue. Object identification can essentially encompass the naming, addressing, or both of an asset or device. In the web, the identification of a resource that represents some form of information has been achieved by the development of the universal resource identifier (URI), which is a global agreement on the identification of a particular resource based on specified schemes. In IoT, similar to the Internet and the web, objects need to have common naming and addressing schemes and also discovery services to enable global reference and access to them. In this section, we review common identification, naming, and addressing schemes and frameworks that can contribute to designing a naming and addressing scheme for IoT/WoT.

Table 6.2 Unified IoT Data Standard Based on Existing Data Formats and Protocols

IoT Standards Matrix ©Zhou	*Existing Data formats and Protocols*		*Unified New IoT Standards*		
			Goals	*Necessity*	*Feasibility*
Application Layer (M)	Data Formats	BITXML, EXDL, MDMP, M2MXML, NGTP, oBIX, oMIX, ONS/PML, OPC, SyncML, WMMP, etc.	Create new, unified, open, cross-sector, usable data standards including formats, exchange protocols, processing and modeling frameworks.	**High** Enable easier integration and interoperability.	**Medium** Enhanced HTML and ebXML-like standards, hard to create a unified data format due to domain differences.
	Software Framework	ArchestrA, CoAP, DRM, ECF, ezM2M, HYDRA, IDM, MDM, OSGi, PaaS, RESTful, SaaS, Sedona, SOA, SODA, SOAP, etc.	Data standards compliant SaaS/ PaaS 3-tiered platform middleware, support new paradigms such as DRM.	**High** Enable easier integration, new MAI paradigm, etc.	**High** Enhanced 3-tiered application servers, OSGi middleware for server-side.

continued

Table 6.2 (continued) Unified IoT Data Standard Based on Existing Data Formats and Protocols

IoT Standards Matrix ©Zhou	*Existing Data formats and Protocols*		*Unified New IoT Standards*		
			Goals	*Necessity*	*Feasibility*
Transmission Layer Protocols (C)	Wired Long Distance	IP(TCP/UDP/HIP), IP over Everything/ Everything over IP, Ethernet, IPv6, ATM, Frame Relay, SDH, FDDI, Fiber Channel, ISDN, SS7, PSTN, VPN, VoIP, Cable/xDSL, etc.	"3-network" convergence, all-IP networks, IPv6 should be leveraged for IoT applications, existing networks OK for most IoT applications.	**Medium** MTC support enhancements and optimizations.	**Medium** It takes time for all-IP, IPv6 to prevail.

	Wired Short Range	ANSI C12.18, AS-i, BACNet, CanBus, CC-Link, ControlNet, Dali, DeviceNet, DF-1, DLMS/IEC 62056, Dupline, FF, FlexRay, HART, HomePNA, IEC 61107, InterBus, LIN, LonWorks, KNX, ModBus, MOST, MTConnect, P-Net, ProfiBus, SwiftNet, Vnet/IP, WorldFIP, CC-Link, PLC, Industrial Ethernet, RS232, RS485, VAN, etc.	Ruggedness enhancements, few new protocols are required, no need to reinvent the wheel.	**Low** Few or no new protocols required.	**Low** Few or no new protocols required.

continued

Table 6.2 (continued) Unified IoT Data Standard Based on Existing Data Formats and Protocols

IoT Standards Matrix ©Zhou	*Existing Data formats and Protocols*		*Unified New IoT Standards*		
			Goals	*Necessity*	*Feasibility*
	Wireless Long Distance	2G: GSM, CDMA, etc.; 3G:WCDMA,EV-DO,HSUPA, EV-DOrA, UMTS, etc.; 2.5G: GPRS, EDGE,HSCSD, etc.; 4G:EV-DOrB, LTE, WiMAX, UMB/UWB, TD-SCDMA, etc. Satellite M2M, GPS, etc.	All-IP, Mobile IP, etc., helpful but not required, MTC enhancements for low bandwidth, low latency IoT applications, backend BOSS system enhancements.	**Medium** Dedicated packet switch MTC network helpful but not required.	**Medium** Few or no new protocols required, optimization focus.

	Wireless Short Range	Bluetooth, BSN, DECT, DSAH 7, EDACS, EnOcean, HyperLan, HyperMAN, 6LoWPAN, HomeRF, HomeIR, InfiNET, Insteon, IrDA, IRIG, ISA 100.11a, LMDS, NFC, OpenSky, OSIAN, RFID, TETRA, TransferJet, WAVE, Wavenis, WiFi/WAPI, WirelessHART, Zigbee, Z-Wave, etc.	Few new protocols required, focus should be on embedded OS or middleware, TinyOS, MagnetOS, Contiki, Mantis, SINA, SensorWare, etc.	**Medium** Enhancements on embedded OS and middleware.	**Low** Few or no new protocols needed, leverage existing protocols.
	Sensor Layer (D)	TEDS/IEEE 1451, CBRN, TransducerML, SensorML, IRIG, EXDL/CAP, AutomationML, OpenPLC XML, EDDL, FDT, CANOpen, etc.	Optimized and minimized version of application layer XML data standards, supported by embedded OS and middleware. Universal OSGi middleware for device-side hardware.	**High** Enable easier integration and interoperability.	**Medium** Minimized ebXML-like standards, it's hard to create a unified standard due to small footprint.

The ubiquitous ID (uID) framework [124] was developed in Japan. uID or ucode is the identification number assigned to individual objects. The ucode is a 128-bit fixed-length identifier system. Moreover, a mechanism to extend the ucode length in units of 128 bits has been prepared to meet the future demands so that codes longer than 128 bits also can be defined.

In the field of RFIDs, EPCglobal [51] has promoted the adoption and standardization of electronic product code (EPC), which has been used to uniquely identify RFID tags. It is based on the URI model. ID@URI, developed by the DIALOG research project [125], is another identification model that takes the same properties of the EPC/ONS standard but can be manifested in bar codes as well. EPC (recognized in the United States and Europe) is a competing standard of uID (used in Japan), affected by national or regional interests, so compatibility and interoperability is always an issue politically instead of technologically.

In the mobile telecoms domain, the international mobile equipment identity (IMEI) [126] provides a means for unique identification of mobile phones. IMEI is formed through a set of digits that represent the manufacturer, the unit itself, and the software installed on it. IMSI conforms to the recommendation of ITU-T E.212 stored in the SIM card and often used as a key in the home location register. For public switched telephone network (PSTN), the operator has the possibility to identify uniquely the resource with the E.163/E.164 addresses (a.k.a., telephone numbers). Besides, operators provide the mobile subscriber ISDN number (MSISDN) following the ITU-T recommendation E.164. This unique number is used for routing calls in the operator networks.

The following unique ID schemes refer to addresses and names of electronic objects at various levels of the OSI stack along with their related protocols: MAC address, IP address on the Internet, e-mail address, uniform resource name (URN), URI, URL, and others. IP address is certainly a straightforward unique ID scheme; however, on January 19, 2010, the Numbers Resource Organization, the entity tasked with protecting the unallocated pool of remaining IPv4 Internet addresses, issued a statement indicating that less than 10 percent of IPv4 addresses remain unallocated. Obviously, if millions to hundreds of millions of new devices are going to be networked in an Internet of Things in the coming years, this shortage of IPv4 addresses poses a challenge, particularly for countries outside of North America that were allocated comparatively fewer IPv4 addresses to begin with. The long-term solution is IPv6, which enables orders of magnitude larger numbers of available IP addresses. Most mobile network operators (MNOs) are in the planning stages for this transition to IPv6 or have already made the transition. M2M-optimized mobile infrastructure can help with the transition by future-proofing applications through the use of techniques such as IPv6 tunneling over IPv4. Essentially, this capability would enable remote M2M devices to use native IPv6 addresses that are translated to IPv4.

In the software domain, the UUID (universal unique identifier, as shown in Figure 6.16) was proposed in the early 1980s. It's an identifier standard used in software construction, standardized by the Open Software Foundation (later called the Open Group) as part of the Distributed Computing Environment. In 1996, it became part of ISO/IEC 11578 documents, and more recently documented in ITU-T Rec. X.667 | ISO/IEC 9834-8:2005. The IETF has published Standards Track

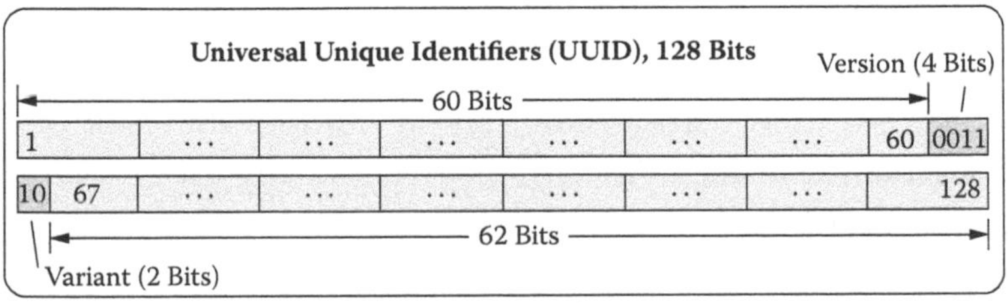

Figure 6.16 Structure of UUID.

RFC 4122, which is technically equivalent with ITU-T Rec. X.667 | ISO/IEC 9834-8. The intent of UUIDs is to enable distributed systems to uniquely identify information without significant central coordination. UUIDs are widely used in distributed middleware such as Tuxedo (the author used to work in the team), CORBA, and JavaEE. One widespread use of this standard is in Microsoft's globally unique identifiers (GUIDs, a different name for UUID). Other significant uses include Linux's ext2/ext3 file systems, GNOME, KDE, and Mac OS X, all of which use implementations derived from the UUID library found in the e2fsprogs (Ext2 file systems utilities) package. UUID was also used in the Bluetooth standard.

The ASN.1 project was established in February 2001 by ITU-T Study Group 7 to assist existing users of ASN.1 within and outside of ITU-T, and to promote the use of ASN.1 across a wide range of industries and standards bodies. Since September 2001, the responsibility for the ASN.1 project resides with Study Group 17 and the project now also encompasses object identifiers (OIDs) and registration authorities.

In an open and international world such as the one of telecommunications and information technologies, one often needs to reference an object in a unique and universal way. Many standards define certain objects for which unambiguous identification is required. This is achieved by assigning OID to an object in a way that makes the assignment available to interested parties. It is carried out by a registration authority. The naming structure of OID is a tree structure that allows the identification of objects in a local or international context, without being limited by the registration authority or by the number of objects they can register (Figure 6.17). Each new node is associated with a name and a number that will be used for data transfers. An OID is semantically an ordered list of object identifier components, for example, {joint-iso-itu-t(2) ds(5) attributeType(4)distinguishedName(49)}, or for short 2.5.4.49. OID and ASN.1 is also widely used in X.400/X.500,

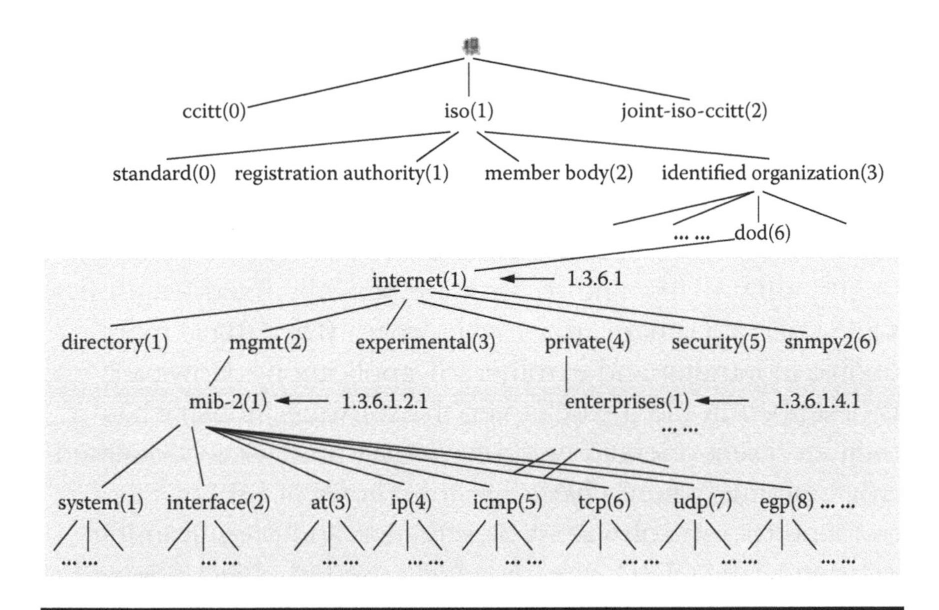

Figure 6.17 Hierarchy of OID.

H.323, SNMP, and in many wired or wireless network protocols such as UMTS, etc.

OID is a flexible, extensible framework. It can also be used together with other ID schemes such as UUID, uID, EPC, and others. For example, the member body number in China as a country is 156. An IoT object's locally (such as Beijing with a number 10) unique ID such as 66666666 can be prefixed with 1.2.156 to form a globally unique ID 1.2.156.10.66666666, just like the phone number prefixed with a country number.

OID is a good identification candidate for IoT objects considering it's a mature scheme and supported by both ISO and ITU. However, it's a bit complex to use (since it is part of ASN.1 involving registration processes etc.) compared with other schemes such as UUID, EPC, or uID. Considering EPC and uID are not compatible with each other due to the aforementioned reasons, UUID is a widely accepted scheme used in distributed environments including IoT (which is a distributed system by itself) and it already exists in many software systems, so UUID is a better scheme for IoT (not necessary on the Internet), especially WoT applications.

Compared with creating a unified, cross-segment, overarching data standard for WoT, it is possible and a must to create a global unique identification. The hurdles are more about interest considerations of related parties rather than technological difficulty.

EPC, uID, UUID, and so forth are basically fixed-length IDs, while OID and others are variable-length IDs. OID is more flexible in intranet and extranet IoT applications. However, as described in the next chapter, the software industry has been an object-oriented world for a long time. Object-oriented programming (where *objects* are the *Things* of IoT) has a profound root in software representation and programming, so using UUID/GUID, already widely used in object-oriented systems, as the identification of Things would be a breeze to transition from object-oriented to real-world objects and the integration of IoT systems with existing IT systems.

6.4 Summary

In this chapter, we talked about the difference between WoT and IoT as well as the web and the Internet. To build WoT, the standardization of communication protocols, especially data formats, plays a crucial and important role as evidenced by the invention and dominance of the HTML/HTTP standard. One of the main value propositions or suggestions to the IoT industry in this book is to focus on protocol standardization, especially data format standardization, instead of standardization on other layers of the value chain, such as creating or modifying existing communication protocols such as Zigbee and others.

Various kinds of existing and emerging protocols in all four pillar segments are investigated and analyzed to support the proposition. However, the feasibility of creating a unified XML data format including a global identification scheme of objects for all IoT applications that cover the four pillar segments is still under investigation by some of the IoT projects worldwide, particularly in Europe. Issues existing in current standardization efforts are also discussed.

In the next chapter, we will talk about existing IoT architectures and the unified architectural framework for IoT.

Chapter 7

Architecture Standardization for WoT

7.1 Platform Middleware for WoT

Current markets of the Internet of Things (IoT) and Web of Things (WoT) are highly fragmented. Various vertical WoT/IoT solutions have been designed independently and separately for different applications, which inevitably impacts or even impedes large-scale WoT deployment. A unified, horizontal, standards-based platform is the key to consolidate the fragmentation.

We talked about communication middleware for IoT in Chapter 5. Communication middleware and platform middleware are closely related to and sometimes tightly integrated with each other. However, there are differences between them. We will talk about platform middleware (also called application frameworks, or sometimes, directly, the three-tiered application server) for IoT in this chapter, especially frameworks at the application level, or at the "M" level of the DCM (direct, change, manage) value chain. One main goal of platform middleware is to bring the IoT applications (including Intranet of Things and Extranet of Things) to the World Wide Web, so we will use the term *Web of Things* more in this chapter.

According to the WoT/IoT vision, everyday objects such as domestic appliances, actuators, and embedded systems of any kind in the near future will be connected with each other and with the Internet. These will form a distributed network with sensing capabilities that will allow unprecedented market opportunities, spurring new services, including energy monitoring and control of homes, buildings, industrial processes, and so forth. In this chapter, we concentrate on the actual implementation of the multitiered application-level technologies.

An interesting observation is that many software architectures and technologies have long before used the term *object* in many modeling methodologies such as the well-known object-oriented design, object-oriented software engineering and programming, CORBA (common object request broker architecture), DOM (document object model), POJO (plain old Java object), COM (component object model) and DCOM (distributed COM), OPC (object linking and embedding for process control), OID (object identification), SOAP (simple object access protocol), JSON (JavaScript object notation), and so on. The entire software industry is already an object-oriented world, as shown in Figure 7.1. The representation and programming of objects has a profound supporting base in the software world.

Now that IoT/WoT brings the real-world objects into the game, there must be many natural fits for mapping the IoT

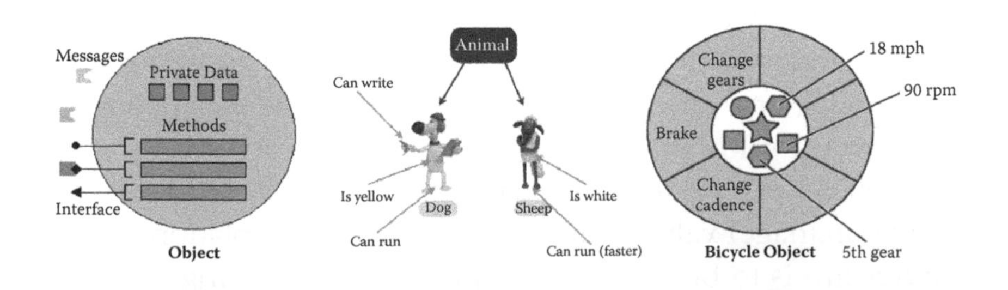

Figure 7.1 Object-oriented and real-world object programming.

objects to software objects. The transition from object-oriented to real-world objects programming is a natural one. In fact, the Mango open-source software platform for IoT and machine-to-machine (M2M) has found a natural fit between JSON and M2M applications and used the technology in its products.

7.1.1 Standards for M2M

The European Telecommunications Standards Institute (ETSI's) Global Standards Collaboration (GSC) M2M Standardization Task Force (MSTF) considers as M2M any automated data exchange between machines including virtual machines such as software applications without or with limited human intervention as described in the previous two chapters. The Technical Committee's overall objective is creating open standards for M2M communications to foster the creation of a future network of objects and services so that already-existing and rapidly growing M2M businesses based on vertical applications using a multitude of technical solutions and diverse standards can be turned into interoperable M2M services and networked businesses. An ETSI M2M architecture diagram (http://www.telit.com/img/images%20market%20intelligence/m2msystemarchitecture.jpg) shows the high-level approach to invert the pipes. Vertical proprietary applications shall be substituted by a horizontal architecture, wherein applications share common infrastructure, environments, and network elements. An M2M system described by clearly structured and specified network transitions, software and hardware interfaces, protocols, frameworks, and so forth shall ensure the interoperability of all system elements. The Technical Committee's work is based on the general guideline of using existing standardized systems and elements. It evaluates them according to M2M requirements, filling gaps as necessary by either enhancing existing standards or producing supplemental ones.

The key elements of the ETSI M2M architecture are described below:

- M2M device: A device capable of replying to requests or transmitting data contained within those devices autonomously.
- M2M area network (MAN): A network providing connectivity between M2M devices and gateways. Examples of M2M area networks include personal area network technologies such as (wireless) IEEE 802.15, short range devices (SRD), UWB, ZigBee, Bluetooth, and others, and (wired) CanBus, Modbus, KNX, LonWorks, PLC (Power Line Communication), and others.
- M2M gateway: The use of M2M capabilities to ensure that M2M devices interwork and interconnect to the communications networks.
- M2M communications networks: Communications networks between M2M gateways and M2M applications servers. They can be further broken down into access, transport, and core networks. Examples include xDSL, PLC, satellite, LTE, GERAN, UTRAN, eUTRAN, W-LAN, WiMAX, and others.
- M2M application server: The middleware layer where data goes through the various application services and is used by the specific business-processing engines.

The M2M platform middleware normally covers the layers from M2M gateway to the M2M application server. As an example, Actility (Active Utility) offers core infrastructure components and software enabling mass-scale, mission-critical applications of the Internet of Things with a specific focus on smart grid applications. Actility designed ThingPark® (http://www.actility.com/thingpark), a hosting infrastructure and marketplace for M2M/IoT applications managing data flows based on open architectures such as ETSI M2M. Realizing the IoT was still missing an architectural framework capable of handling such scale and truly enabling interoperability, Actility became a contributor to the ETSI M2M architecture-level standard and decided to develop an open-source implementation for IoT gateway developers, embedding all major existing M2M, sensors, and automation protocols. All referenced hardware platforms support OSGi (Open Services Gateway initiative framework) execution of ThingLets® and are remotely configured by the ThingPark infrastructure.

ComSoc Communities, an IEEE (Institute of Electrical and Electronics Engineers) collaboration of industry professionals, reports that United Parcel Service and Cinterion with TZ Medical have adopted ETSI standardized M2M applications, each for different purposes. UPS was able to achieve a 3.3 percent reduction in the amount of fuel consumed per package in the United States, and to reduce engine idling time by 15.4 percent in 2010. The system uses M2M technology comparable to iMetrik's to monitor and wirelessly report vehicle performance and driving habits, and route information to a central location. Cinterion and TZ Medical collaborated to launch a new heart-monitoring device to detect cardiac abnormalities in patients and communicate the diagnostic data to physicians through mobile networks and the Internet. Designated caregivers can track patient data at any time, from any place, to make treatment decisions.

The existing OMA (Open Mobile Alliance) and its M2M Task Force is producing a white paper that identifies M2M standards gaps and recommendations for OMA actions. Several OMA standards provide building blocks that map into the ETSI M2M framework:

- Device management can provide ETSI's remote entity management service
- Gateway management object fulfills some ETSI gateway service requirements
- Firmware updates, software updates, provisioning, diagnostics, and monitoring
- Converged personal network services maps into ETSI M2M area network

- Reachability, address mapping, inter/intra-area-network messaging, service publication and discovery
- Some OMA enablers (e.g., location) support services that can be used in M2M applications

An OMA graphic [235] shows how OMA enablers map to ETSI generic M2M framework (that applies to all M2M applications) using telematics application components as examples. OMA Converged Personal Network Services maps to ETSI M2M Area Network, GwMO maps to ETSI Gateway, OMA Device Management maps to ETSI Remote Entity Management, and so forth.

The Car-to-Car Communication Consortium, originally initiated by European vehicle manufacturers and now open to other partners, aims at providing a means to improve road safety by defining vehicle-to-vehicle and vehicle-to-infrastructure communications mechanisms, as described in Chapter 2. Multiple roadside units are deployed along roadsides. These devices communicate and act as a gateway for vehicular on-board units. Vehicle identifiers can use Internet Protocol version 6 (IPv6) addresses for the vehicles' on-board units. It is enhanced to enable geographical routing, which allows a sender's application to issue a message targeting recipients located in a certain geographical area. Also, International Organization for Standardization (ISO) TC204 WG16 (working together with ETSI Technical Committee Intelligent Transport Systems [TC ITS]) is drafting a series of standards under the acronym CALM (continuous communications air interface for long and medium range). The objective of this standard is to provide an architecture framework and a set of protocols for vehicle-roadside communications that separate applications from the communications media.

7.1.2 Frameworks for WSN

The Open Geospatial Consortium, Sensor Web Enablement (OGC SWE) standardization effort is intended to be a

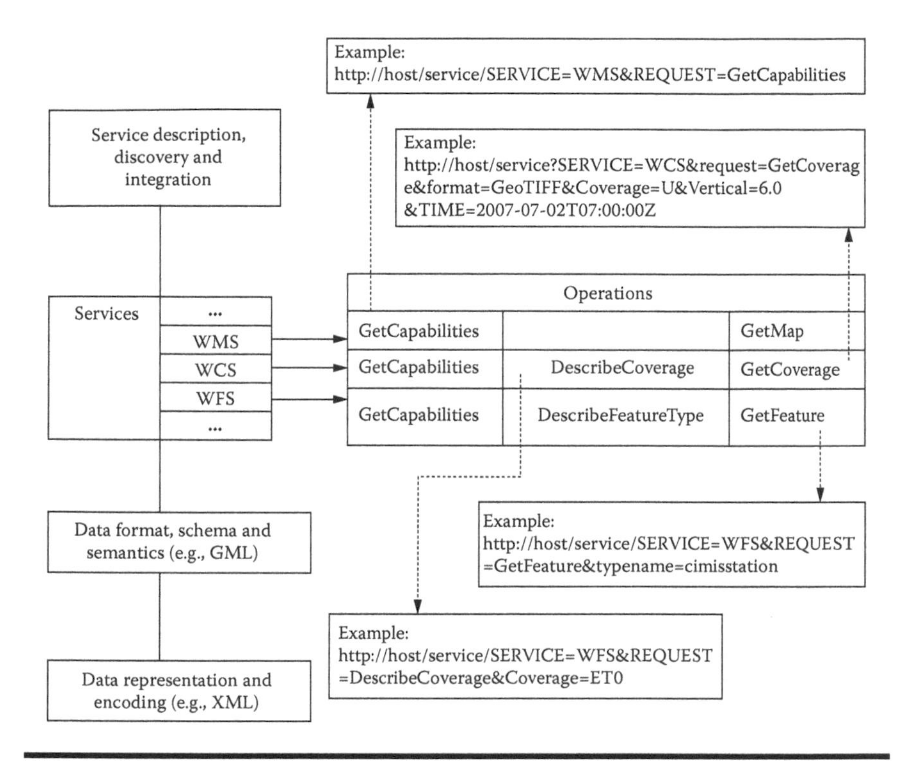

Figure 7.2 OGC services, operations, and example calls (indicated by dotted lines) for web map service (WMS), web coverage service (WCS), and web feature service (WFS). (From Michael Gertz, Carlos Rueda, and Jianting Zhang, "Interoperability and Data Integration in the Geosciences," in Arie Shoshani and Doron Rotem (Eds.), *Scientific Data Management: Challenges, Technology, and Deployment*, Boca Raton, FL: CRC Press, 2010.)

revolutionary approach for exploiting web-connected sensors such as flood gauges, air pollution monitors, satellite-borne earth-imaging devices, and so forth. The goal of SWE is creation of web-based sensor networks to make all sensors and repositories of sensor data discoverable, accessible, and where applicable, controllable via the World Wide Web (Figure 7.2).

SWE standards are developed and maintained by OGC members who participate in the OGC Technical Committee's SWE Working Group. SWE is a suite of standard encodings and web services that enable [106] the following:

- Discovery of sensors, processes, and observations
- Tasking of sensors or models
- Access to observations and observation streams
- Publish–subscribe capabilities for alerts
- Robust sensor system and process descriptions

The following web service specifications have been produced by the OGC SWE Working Group (in addition to the encoding specifications described in Chapter 6):

- Sensor observation service—standard web interface for accessing observations
- Sensor planning service—standard web interface for tasking sensor systems and model and requesting acquisitions
- Sensor alert service—standard web interface for publishing and subscribing to sensor alerts
- Web notification service—standard web interface for asynchronous notification

The USN (Ubiquitous Sensor Networks) standardization of ITU-T is another effort being carried out under the auspices of the Next-Generation Network Global Standards Initiative (NGN-GSI). USN is a conceptual network or framework built over existing physical networks that makes use of sensed data and provide knowledge services. Its main components are as follows:

- USN applications and services platform: technology framework to enable the effective use of a USN in a given application or service
- USN middleware: including functionalities for sensor network management and connectivity, event processing, sensor data mining, and so forth
- Network infrastructure: mainly based on NGNs, USN is not a physical network but rather a conceptual network making use of existing networks

- USN gateway: A node that interconnects sensor networks with other networks
- Sensor network: Network of interconnected sensor nodes (IP-based nodes with direct connection to NGN, non-IP-based nodes connected to NGN via gateways and others)

7.1.3 Standards for SCADA

ISO 16100-1:2009, one of the components of ISO 16100 standard for industrial automation systems and controls–IT convergence integration, specifies a framework for the interoperability of a set of software products used in the manufacturing domain and to facilitate its integration into a manufacturing application. This framework addresses information exchange models, software object models, interfaces, services, protocols, capability profiles, and conformance test methods.

ANSI/ISA-95 is an international standard for developing an automated interface between enterprise and control systems. This standard has been developed for global manufacturers. It was developed to be applied in all industries and in all sorts of processes, like batch processes, continuous and repetitive processes aiming to reduce cost, and risk and errors associated with implementing interfaces between enterprise and production control systems. It continues to be developed and refined by the Instrumentation, Systems, and Automation Society (IAS) in collaboration with major vendors of ERP and MES solutions around the world.

The objectives of ISA-95 are to provide consistent terminology that is a foundation for supplier and manufacturer communication, to provide consistent information models, and to provide a consistent operations model as a foundation for clarifying application functionality and how information is to be used.

The five parts of the ISA-95 standard are as follows:

- ANSI/ISA-95.00.01-2000, Enterprise-Control System Integration, Part 1: Models and Terminology
- ANSI/ISA-95.00.02-2001, Enterprise-Control System Integration, Part 2: Object Model Attributes
- ANSI/ISA-95.00.03-2005, Enterprise-Control System Integration, Part 3: Models of Manufacturing Operations Management
- ISA-95.04, Object Models & Attributes, Part 4: Object models and attributes for Manufacturing Operations Management
- ISA-95.05, B2M Transactions, Part 5: Business to Manufacturing Transactions

OPC Unified Architecture [118] brings two elementary innovations into the OPC world (Figure 7.3). On the one hand, the Microsoft Windows–specific protocol DCOM is replaced by open, platform-independent protocols with integrated security mechanisms. On the other, the proven OPC features, such as data access, alarms and events, and historical data access, are

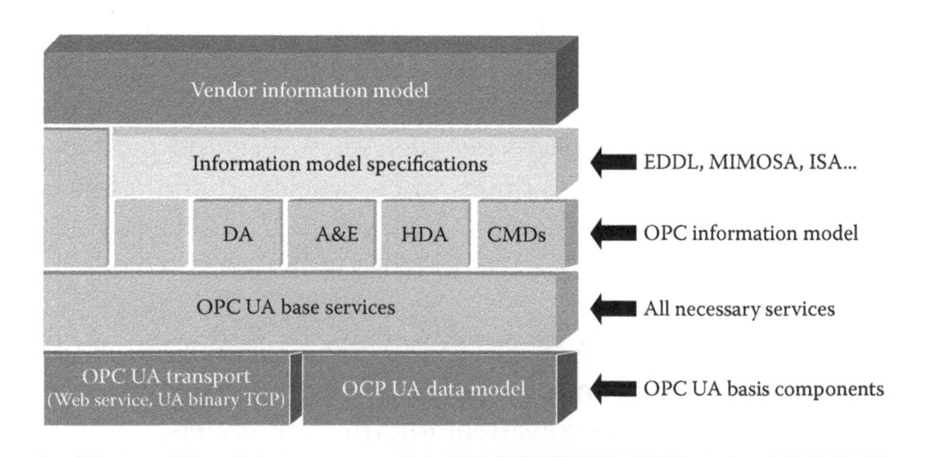

Figure 7.3 OPC unified architecture. (From Tuan Dang and Renaud Aubin, "OPC UA," in J. David Irwin (Ed.), *Industrial Communication Systems*, Boca Raton, FL: CRC Press, 1998.)

summarized in an object-oriented model and supplemented by new and powerful features, such as methods and type systems. As a result, not only can the OPC interface be directly integrated into systems on arbitrary platforms with different programming languages, but arbitrary complex systems can also be described completely with OPC UA. It can be implemented with JavaEE, Microsoft.NET, or C, eliminating the need to use a Microsoft Windows–based platform of earlier OPC versions. UA combines the functionality of the existing OPC interfaces with new technologies such as XML (extensible markup language) and web services to deliver higher-level MES and ERP support. The OPC Foundation and the MTConnect Institute announced their cooperation to ensure interoperability and consistency between the two standards in 2010.

ISA-95 and ISA-88 standards define information models for production control systems, batches, and MES. Their mapping to OPC UA is planned as an ISA-95 companion standard. Oracle provides ISA-95 standard–based integration capability between MES. Many other IT software vendors also provide ISA-95/88– and OPC UA–compliant products. For example, Wonderware's ArchestrA™ Platform provides support (http://global.wonderware.com/EN/PDF%20Library/Enterprise_Integration_Application_White_Paper.pdf) for open information standards ISA-95 and the message structures defined in ISA-95's B2MML (business-to-manufacturing markup language) messages.

On a different front, the SmartProducts [108] consortium aimed at demonstrating practical research that resulted in a platform that supports stand-alone or integrated, context-aware products for a range of application scenarios. It developed a scientific and technological basis for building smart products with embedded "proactive knowledge." Details about the different components of the middleware platform can be found in the following categories:

- Interaction: components for supporting the interaction between user and smart products
- Communication: components for supporting the information exchange and cooperation between different smart products
- Context: components for sensing, processing, and distributing context information
- Proactive knowledge base: components for handling the knowledge of a smart product
- Secure distributed storage: components for storing knowledge of a smart product in a secure and distributed way
- Tools: tools for developing smart products, such as for automatically extracting relevant information from manuals, editors

7.1.4 Extensions on RFID Standards

The EPCglobal-defined radio-frequency identification (RFID) architecture and frameworks are probably the most comprehensive and complete standards among the four pillar segments of IoT. Again, we will not talk about the EPCglobal effort because it was described in previous chapters of the book.

Many efforts to define IoT as described in Chapter 1 are of RFID origin. CASAGRAS (Coordination and Support Action for Global RFID-related Activities and Standardization) [110] was one of them, an FP7 project that ended in 2009 after 18 months (FP7 is Seventh Framework Programme, an initiative that bundles all research-related European Union initiatives together under a common roof playing a crucial role in reaching the goals of growth, competitiveness, and employment). The goal of CASAGRAS was to provide a framework of foundation studies to assist the European Commission and the global community in defining and accommodating international issues and developments concerning RFID with particular reference to the emerging Internet of Things. It seems that nothing particularly useful or better than EPCglobal was generated by this effort.

CASAGRAS2 started in June 2010 and ended in June 2012. The consortium consists of partners from Europe, the United States, China, Japan, Brazil, and Korea. The stated goal is to address the key international issues that are important in providing the foundations and cooperation necessary for realizing the Internet of Things as a global initiative.

BRIDGE [111] (Building Radio-frequency IDentification solutions for the Global Environment) is a European Union–funded three-year integrated project addressing ways to resolve the barriers to the implementation of RFID in Europe, based upon GS1 EPCglobal standards, by extending the EPC network architecture. One of the core aspects of BRIDGE related to IOT-A lies in the Discovery Service, which manages the exchange of RFID and aggregated information between nodes.

The Cross UBiQuitous Platform (CUBIQ) [112] project, in which nine organizations in Japan participate, aims to develop a common platform that facilitates the development of context-aware applications. The idea is to provide an integrated horizontal platform that offers unified data access, processing, and service federation on top of existing, heterogeneous IoT-architecture-based ubiquitous services. A unified data model was defined using USDL (universal service definition language). The CUBIQ architecture consists of three layers (and serverless real-time location search with RFID tags is an application example of the CUBIQ project)

- Mobile terminals with RFID tag reader collect RFID tag info and record location.
- The mobile terminals are connected via the core CUBIQ infrastructure and share RFID tag information.
- Observers can search RFID tag information to estimate the location of target person.

7.2 Unified Multitier WoT Architecture

Apart from the standard efforts such as the ETSI, 3rd Generation Partnership Project (3GPP), and Open Mobile Alliance (OMA), many research projects and industrial products aim to define and build a common middleware platform for WoT/IoT applications.

Niagara^{AX} is a software framework (http://www.neopsis.com/cms/en/solutions/niagara/) product and development environment that solves the challenges associated with building device-to-enterprise applications and distributed Internet-enabled automation systems, which are deployed in over 160,000 installations worldwide. Tridium's Niagara introduced the concept of a software framework that could normalize the data and behavior of diverse devices, regardless of manufacturer or communication protocol, to enable the implementation of seamless, Internet-connected, web-based systems. The data from diverse device systems are transformed into uniform software components. These components form the foundation for building applications to manage and control the devices. The Niagara^{AX} component model goes beyond unifying protocols and data from diverse systems to unifying the entire development environment used to build applications. Here are the Niagara^{AX} highlights:

- New graphics presentation framework and graphic development tool
- Comprehensive library of control objects
- New data archive model and flexible archive destinations
- New alarming capabilities that provides better visualization and user experience
- Reporting and business intelligence supports
- Open driver development toolkit
- Open application programming interfaces (APIs) for developers

FI-WARE of the European Union's Future Internet Core Platform project [236] aims to create a novel service infrastructure, building upon elements called generic enablers that offer reusable and commonly shared functions, making it easier to develop future Internet applications in multiple sectors.

This infrastructure will bring significant and quantifiable improvements in the performance, reliability, and production costs linked to Internet applications, building a true foundation for the future Internet. The reference architecture of the FI-WARE platform is structured along a number of technical chapters:

- Cloud Hosting
- Data/Context Management
- IoT Services Enablement
- Applications/Services Ecosystem and Delivery Framework
- Security
- Interface to Networks and Devices

However, details have not yet been worked out in this FI-WARE project. In the following sections, we will talk about some of the existing technologies that should be leveraged to build a FI-WARE-like unified horizontal framework.

7.2.1 SOA/EAI versus SODA/MAI

As described in the previous chapter, WoT/IoT applications should inherit and enhance the existing data formats and protocols, and the matching software frameworks to build platform middleware for WoT applications. SOAP (simple object access protocol), the successor of XML-RPC, is a protocol framework specification for exchanging structured information in the implementation of web services in computer networks. It relies on XML for its message format, and usually relies on other application layer protocols—most notably hypertext transfer protocol (HTTP), simple mail transfer protocol (SMTP), and Java messaging services (JMS)—for message negotiation and transmission. SOAP, which unified the CORBA, JavaEE, and .NET camps under one umbrella, can form the foundation layer of a web services protocol stack, providing a basic messaging framework upon which web services can be built. SOAP can be a good generic WoT data exchange protocol considering it can tunnel easily over firewalls and proxies of existing infrastructure, among other advantages. Because of the verbose XML format, SOAP can be considerably slower than other competing middleware technologies such as CORBA; however, this may not be an issue when only small messages are sent, which is the case for machine-type communicaiton (MTC) of WoT.

The REST (representational state transfer interface) architecture was developed in parallel with HTTP/1.1, based on the existing design of HTTP/1.0, but it is not limited to the HTTP protocol. RESTful architectures can be based on other application layer protocols if they already provide a rich and uniform vocabulary for applications based on the transfer of meaningful representational state (Figure 7.4). REST is a lightweight SOAP. RESTful applications maximize the use of the preexisting, well-defined interface and other built-in capabilities provided by the chosen network protocol, and minimize the addition of new application-specific features on top of it. REST also attempts to minimize latency and network communication, while at the same time maximizing the independence and scalability of component implementations. The simple

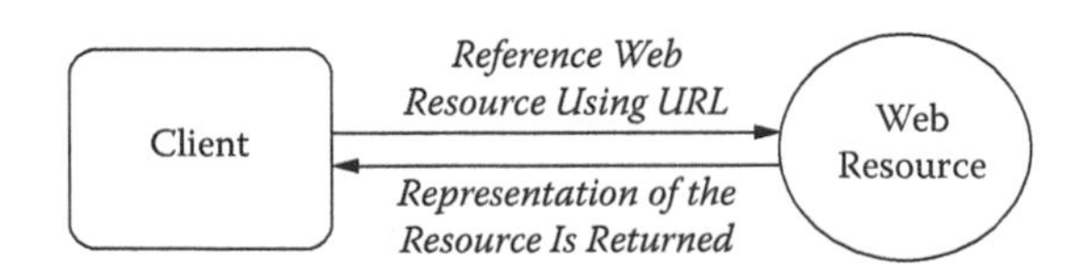

Figure 7.4 RESTful web services. (From Bhavani Thuraisingham, *Secure Semantic Service-Oriented Systems*, New York: Auerbach Publications, 2010.)

semantic of REST and its wide adoption helped provide services that have been reused in other domains like smartphone application. So, REST is a better framework protocol for MTC-based WoT/IoT applications.

For resource-constrained devices, CoAP (constrained application protocol) [127] is a specialized RESTful transfer protocol for use with constrained networks and nodes for M2M applications such as smart energy and building automation. These constrained nodes often have eight-bit microcontrollers with small amounts of ROM and RAM, while networks such as 6LoWPAN often have high packet error rates and a typical throughput of tens of kbit/s. CoAP, similar to SENSEI, provides the REST method/response interaction model between application endpoints, supports built-in resource discovery, and includes key web concepts such as URIs and content types. CoAP easily translates to HTTP for integration with the web while meeting specialized requirements such as multicast support, very low overhead, and simplicity for constrained environments. The Internet Engineering Task Force (IETF) recently approved a new working group called Constrained RESTful Environments (CoRE) based on CoAP's work. This new group aims at specifying a RESTful web service protocol for even the most constrained embedded devices and networks.

SOAP, REST, and CoAP are standard technologies for B2B-like integration of systems on the Internet at the M2M/IoT communication networks layer as described in Section 7.1.1, which should be part of the unified application framework data standards that works over the Internet. There are other standardized technologies such as ESB (enterprise service bus, Figure 7.5) based on MQ (message queue, and MQ_TT for resource constrained networks) and JMS for internal enterprise application integration (EAI) within intranet and extranet. Those technologies can be used for IoT application integration within an intranet or extranet, and they can also be used or extended to work over the Internet.

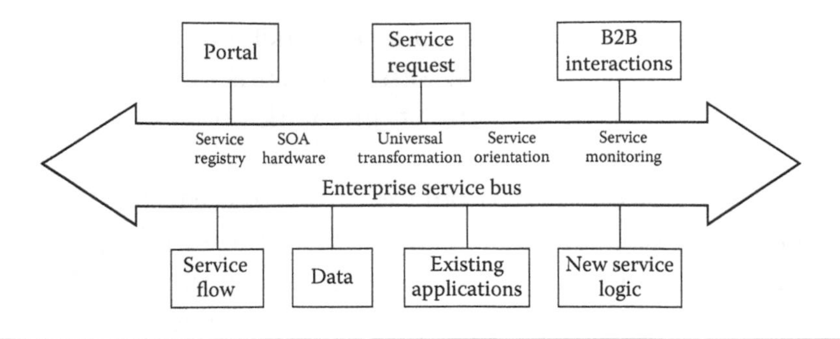

Figure 7.5 Enterprise service bus. (From Yurdaer Doganata, Lev Kozakov, and Mirko Jahn, "Software Architectures for Enterprise Applications," in Mostafa Hashem Sherif (Ed.), *Handbook of Enterprise Integration*, New York: Auerbach Publications, 2010.)

The JCA (Java Connector Architecture) is another good approach for WoT data collection and integration based on connectors or adaptors (the .NET architecture also has adaptors), which has been used in the author's team for the [ez]M2M platform middleware for WoT applications supporting many vertical sectors. JCA is also for internal EAI applications somewhat like OPC for SCADA (supervisory control and data acquisition). JCA-like architecture (http://en.wikipedia.org/wiki/Java_EE_Connector_Architecture) can be used at the M2M/IoT gateway layer. Examples that use adapter/connector architecture include http://www.opengate.es/, http://www.idigi.com/, and so on. Efforts should be spent on making JCA-like architecture work over the Internet if needed.

EAI and B2B seem related but they vary radically in their details. The first is entirely within a single administrative domain. If a new protocol does not work perfectly, it can be ripped out and replaced. In the cross-business environment, ripping it out affects customers, who may have no incentive to upgrade to the new protocol and will be annoyed if it changes constantly. Within a business, demand for a service can be fairly easily judged. On the external interface, demand can spike if a service turns out to be wildly popular with customers. Within a business, security can (to a certain extent) be maintained merely by firing people who abuse it. On the external interface, a much lower level of trust should be extended. Those principles apply to WoT (over the Internet and M2M application integration [74] within an Intranet) applications too.

A service-oriented architecture (SOA) is a set of principles and methodologies for designing and developing software in the form of interoperable services, usually over the Internet. Services comprise unassociated, loosely coupled units of functionality that have no calls to each other embedded in them. SOA requires metadata (unified WoT architecture also needs metadata) in sufficient detail to describe not only the characteristics of the promised services but also the data that drives them. The web services description language typically describes the services, while the SOAP protocol describes the communication protocols. One can, however, implement SOA using any service-based technology, such as REST, CORBA, or Jini, and any programming language.

Web services make functional building blocks accessible over standard Internet protocols independent of platforms and programming languages. These services can represent either new applications or just wrappers around existing legacy systems to make them network enabled. The Web Services Business Process Execution Language is an XML-based execution language that can be used to compose the coarse-grained services into broader services or complete applications. These powerful services are usually orchestrated into processes. The Universal Description Discovery and Integration specification defines a way to publish and discover information about web services as shown in Figure 7.6, which is also a function that WoT applications need.

The combination of the existing SOA (across Internet and extranet) and EAI (intranet) technologies is a good foundation for WoT/IoT applications. EAI can be extended for MAI (M2M

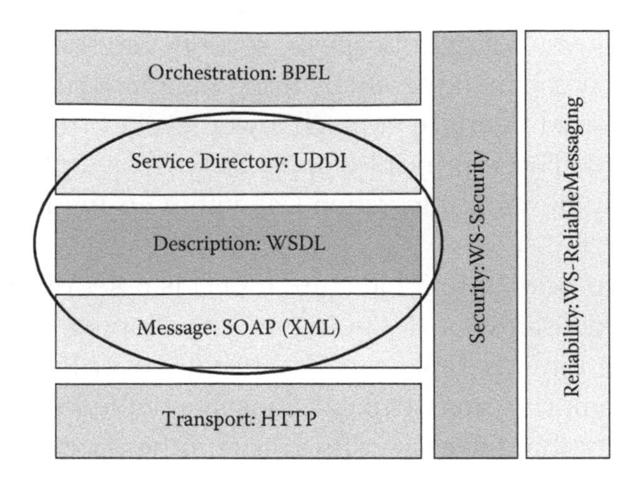

Figure 7.6 B2B technologies. (From Setrag Khoshafian, *Service Oriented Enterprises*, New York: Auerbach Publications, 2007.)

application integration) within an intranet. SOA can be used for WoT/IoT integration over the Internet and extranet.

In fact, a service-oriented device architecture (SODA) is proposed to enable device connection to an SOA (Figure 7.7). The SODA Alliance is an open, customer-driven, broad community chartered to promote consistent integration of the physical world into an SOA network [131]. As described before, developers have connected enterprise services to an ESB using the various web service standards since the advent of XML in 1998. With SODA, which can be based on the OSGi [177] framework described in the next section, developers are able to connect devices to the ESB, and users can access devices in exactly the same manner that they would access any other web service.

The core of the SODA standard is the DDL (device description language) based on XML encodings. DDL classifies devices into three categories: sensors, actuators, and complex devices. Figure 7.8 shows the DDL device model and a sample DDL file of an analog sensor [134]. The ATLAS platform of University of Florida is an implementation of the SODA standard.

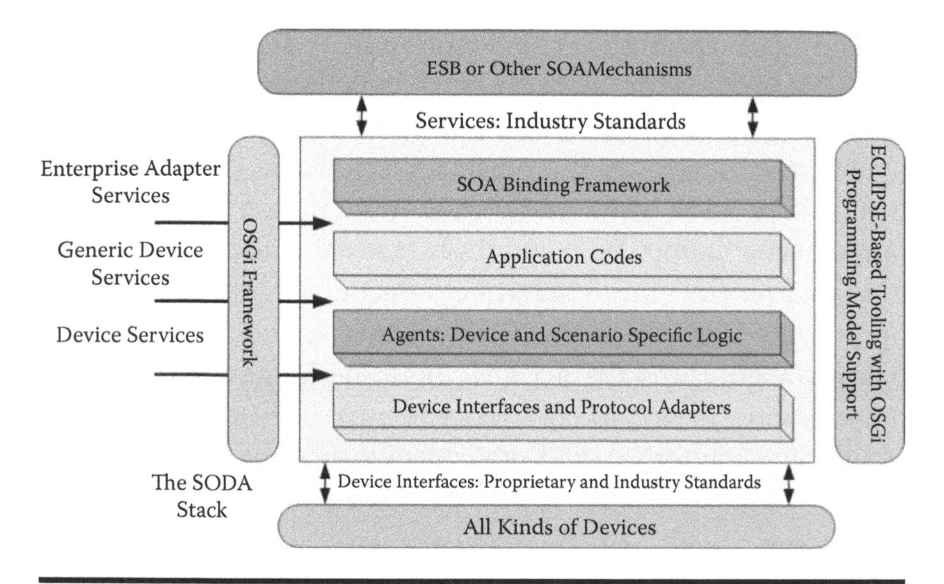

Figure 7.7 SODA architecture.

```
<Sensor>
<Description>…</Description>
<Interface>
<Signal id = "ADC1">…</Signal>
<Reading id = "Temp 1">
<Type>Physical</Type>
<Measurement>Temperature</Measurement>
<Unit>Centigrade</Unit>
<Computation>
<Type>Formula</Type>
<Expression> Temp 1 = (((ADC1/1023 * 3.3)-0.5)*
(1000/10)</Expression>
</Computation>
</Reading>
</Interface>
</Sensor>
```

Figure 7.8 Example of the device description language of SODA.

The Open Healthcare Framework (OHF) is a project based on SODA formed for the purpose of expediting healthcare informatics technology including mHealth [132], or mobile health, a term used for the practice of medicine and public health, supported by mobile devices. The project is composed of extensible frameworks and tools that emphasize the use of existing and emerging standards to encourage interoperable open-source infrastructure, thereby lowering integration barriers. OHF currently provides tools and frameworks for devices and the HL7 (Health Level Seven), IHE (Integrating the Healthcare Enterprise), and other data formats and protocols.

7.2.2 OSGi: The Universal Middleware

The OSGi (Open Services Gateway initiative framework) [121] is a module system and service platform for the Java programming language that implements a complete and dynamic component model. The OSGi Alliance is an open standards organization founded in 1999 that originally specified and continues to maintain the OSGi standard. Using OSGi, applications or components (coming in the form of bundles for deployment) can be remotely installed, started, stopped, updated, and uninstalled without requiring a reboot. Management of Java packages/classes is specified in great detail. Application life cycle management (start, stop, install, etc.) is done via APIs that allow for remote downloading of management policies. The service registry allows bundles to detect the addition of new services or the removal of services, and adapt accordingly. OSGi can have a very small footprint [178] and run on ARM-based devices (e.g., ProSyst OSGi middleware) and operating systems such as Wind River, Android (also on top of a JVM), and so on.

The OSGi specifications have moved beyond the original focus of service gateways and are now used in applications

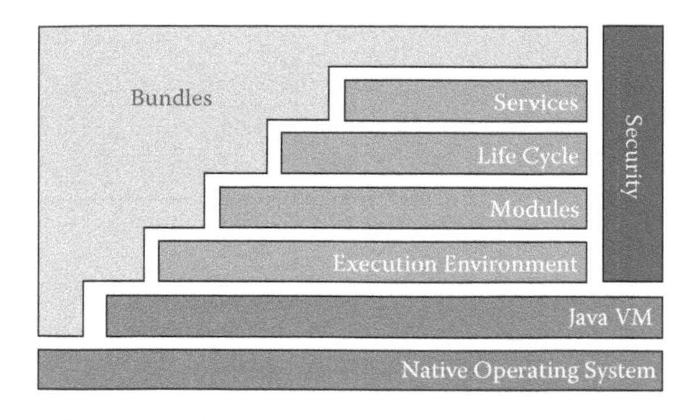

Figure 7.9 OSGi architecture.

ranging from mobile phones to the open-source Eclipse IDE (which dominates the IDE market). Other application areas include automobiles, industrial automation, building automation, PDAs, grid computing, entertainment, fleet management, and application servers (e.g., BEA Systems/Oracle Micro-kernel and SpringSource dm Server). It seems that it is specifically built for IoT/M2M applications considering it can fit in many places in the DCM value chain from device agents to cloud servers. OSGi is a universal middleware [130] and is going to play an important role, as a unified multitiered middleware architecture, in building WoT/IoT applications in many vertical segments (Figure 7.9).

The graphic at http://www.nec.co.jp/techrep/en/journal/g10/n02/100220-122.html depicts an M2M platform and device (gateway and agents) architecture based on the OSGi middleware framework. The ezM2M middleware platform (Figure 7.10) built by the author's team was also migrated from Java application servers to OSGi. OSGi-based platform middleware can provide both a traditional RCP (rich client platform) client-server and a web-based user interface on one platform, which is a very important feature needed by SCADA applications. Another WoT middleware platform built on top of OSGi is the Everyware Software Framework (http://esf.eurotech.com/doc/1.2systemArchitecture.html).

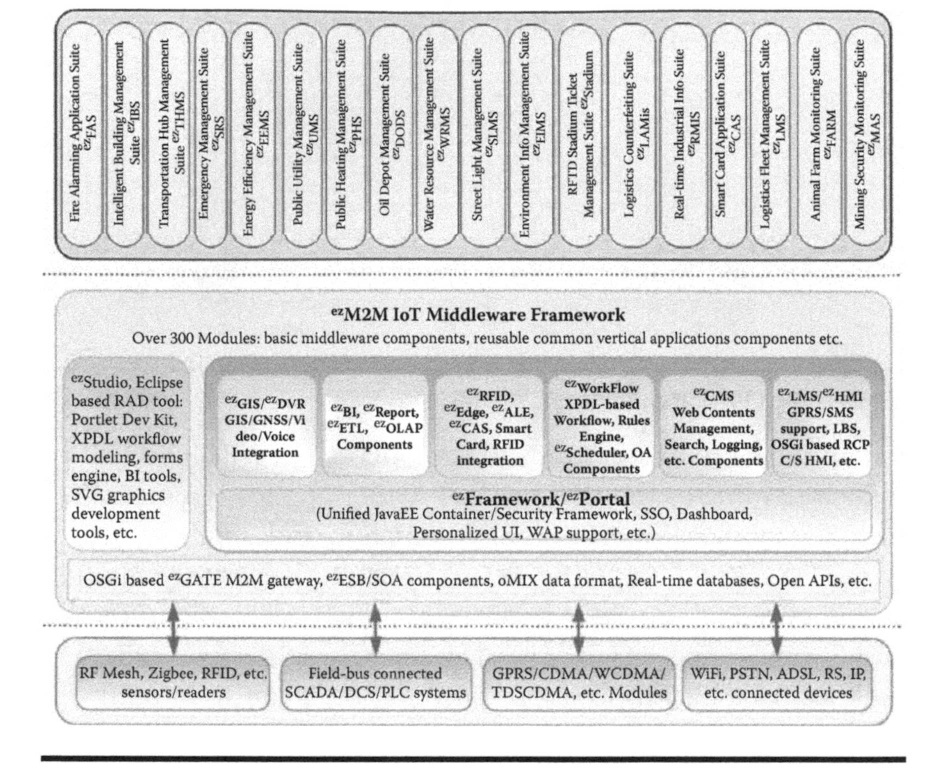

Figure 7.10 **ezM2M platform middleware.**

7.2.3 WoT Framework Based on Data Standards

As discussed in Chapter 5, the platform middleware of WoT can itself be multitiered, just like the multitiered application servers for web applications. In fact, the best realistic approach should be using the existing platform middleware described in the previous two sections to build web-based WoT/IoT applications. An example of such a multitiered architecture is IBM's WebSphere Everyplace (now part of MQ-TT) Device Manager that is based on the three-tier WebSphere application server.

Another three-tiered IoT platform middleware named ezM2M was built by the author's team starting in 2003. It is based on the JavaEE technology and runs on top of three-tiered Java application servers such as JBoss, WebSphere,

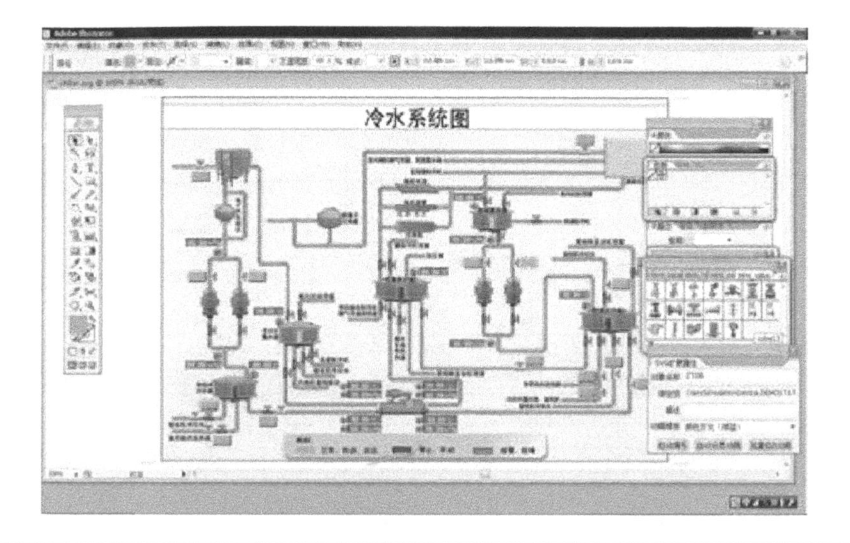

Figure 7.11 ^{ez}Studio IoT application IDE.

WebLogic, and others. More than 18 vertical IoT application suites have been developed on top of ^{ez}M2M, and more than 800 hundred IoT projects have been implemented worldwide, mostly in China.

Figure 7.11 depicts the ^{ez}Studio RAD (rapid application development) environment with an SVG graphic created based on the graphic library that comes with the IDE tool.

Also, a number of new and existing software makers are riding the IoT wave and created middleware for IoT. Some new paradigms have been introduced. For example, Axeda introduced device relation management, OMA proposed MDM (mobile device management), another is intelligent device management, and so forth. An example of MDM implementation is the Fromdistance MDM framework (http://www.empower.com.my/Fromdistance%20MDM.pdf).

Based on the sample middleware platform, a unified multi-tiered IoT middleware can be categorized as having layers as shown in Figure 7.12. The bold outlined blocks are extra tiers that are added to the existing three-tiered application server architecture.

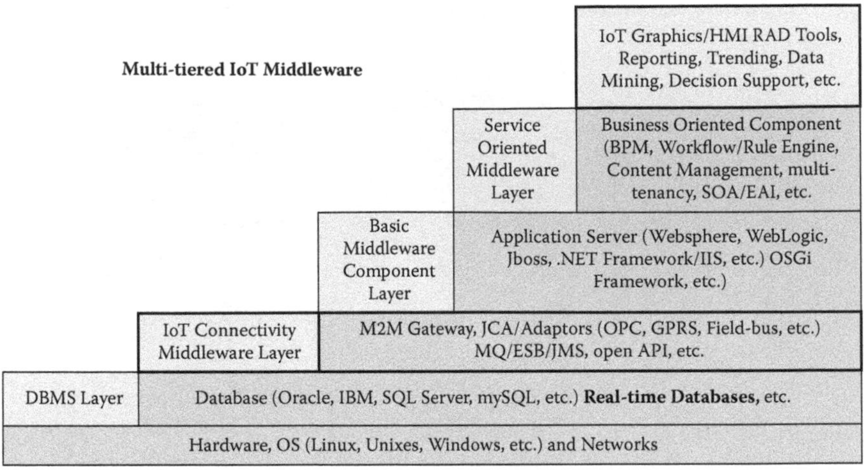

Multi-tiered IoT Middleware

IoT Graphics/HMI RAD Tools, Reporting, Trending, Data Mining, Decision Support, etc.

Service Oriented Middleware Layer | Business Oriented Component (BPM, Workflow/Rule Engine, Content Management, multi-tenancy, SOA/EAI, etc.

Basic Middleware Component Layer | Application Server (Websphere, WebLogic, Jboss, .NET Framework/IIS, etc.) OSGi Framework, etc.)

IoT Connectivity Middleware Layer | M2M Gateway, JCA/Adaptors (OPC, GPRS, Field-bus, etc.) MQ/ESB/JMS, open API, etc.

DBMS Layer | Database (Oracle, IBM, SQL Server, mySQL, etc.) **Real-time Databases,** etc.

Hardware, OS (Linux, Unixes, Windows, etc.) and Networks

Figure 7.12 Multitiered IoT middleware.

The following additional functionalities and tools are added to IoT middleware:

- Drag-and-drop/WYSIWYG (what you see is what you get) graphics and animation development and deployment tools with embedded scripts (Figure 7.13); RAD tools without programming
- BPM/rules engine (no programming required)-based IoT event/alert handling and actions
- M2M gateway, communication adaptors, open and standard API, real-time databases, and so forth

The unified horizontal WoT platform middleware will collect data from the M2M/IoT gateway level and up (or similar level for other WoT pillar systems) as defined by the ETSI/3GPP GSC efforts noted in Section 7.1.1. As an example, the deployment scenario of the M2M software development platform built by InterDigital [137] conforms to ETSI M2M Release 1 standards; however, it doesn't have a unified reference architecture yet.

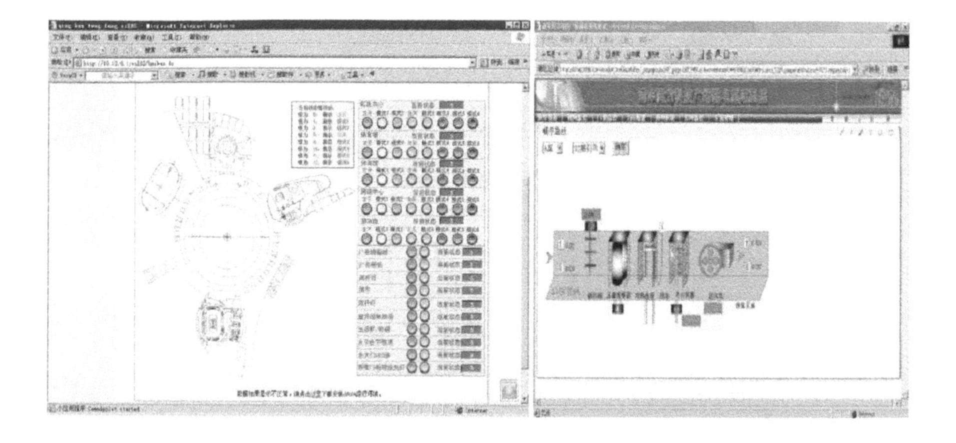

Figure 7.13 IoT graphics with animation.

The IoT-A project of the European Union has just delivered the specification version 1.2 [113] that created a reference architecture. However, this reference architecture considered WSN and RFID only in the unified communication layer, which is not a completely unified IoT architecture as of yet.

Figure 7.14 depicts the unified IoT middleware framework/architecture proposed by the author as a summarization of the previous chapters.

The IoT gateways (that behave as JCA-like IoT adaptors) will be connected to the M2M communication networks, on which the ESB-like (incorporating REST/SOAP functionalities) M2M/IoT communication middleware (we can call it the IoT bus) will reside. The IoT bus will be the IoT integration middleware similar to the SOA/EAI middleware that collect data from all the IoT adaptors, which represent or are the hubs connecting the IoT nodes or subsystems. The IoT platform middleware will finally integrate all the data from the IoT adaptors via the IoT bus. Figure 7.14 is the unified multitiered WoT application architecture framework based on the platform middleware. The multitiered architecture is summarized as follows

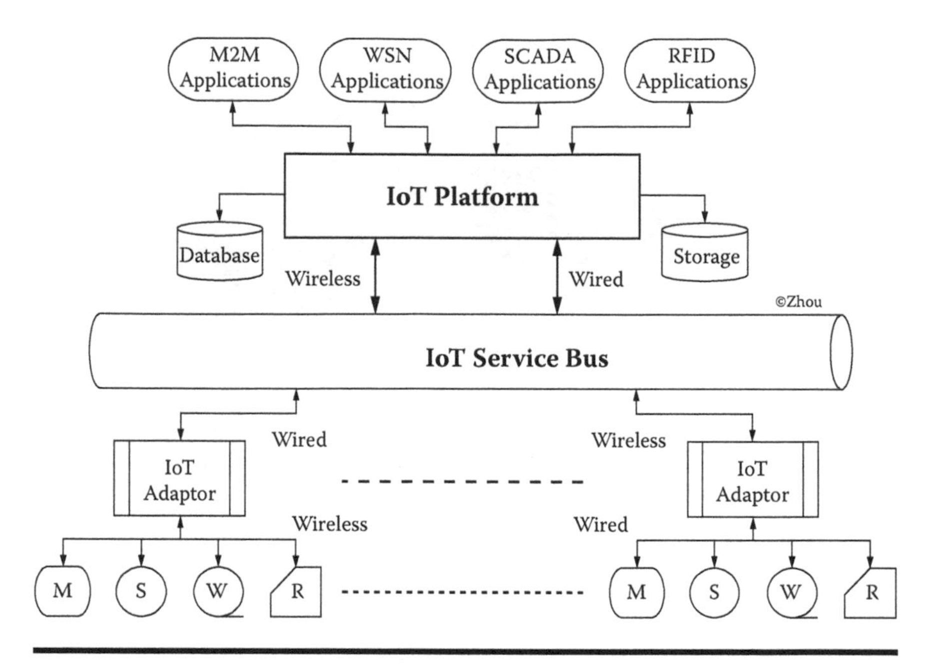

Figure 7.14 The unified IoT middleware framework.

(The ezM2M platform as shown in Figure 7.10 is a reference implementation of this architecture):

- Application framework SES (smart enterprise suite)–like layer for four-pillar applications
- The IoT platform middleware based on application server (container)
- IoT services bus based on ESB (REST/SOAP/MQ/JMS, etc.) and unified XML data format and protocol described in the previous chapter
- IoT adaptor based on the JCA-like adaptor technology for M2M/IoT gateway for device subnet or subsystems
- The back end of IoT hosted by cloud infrastructure and provides IoT cloud (MAI or XaaS) services
- The different devices or sensors of four pillars are connected via the IoT gateways to the IoT bus. They could be mixed or in small birds-of-a-feather groups.

Recent developments on PaaS and SaaS also adopted the approach of extending the application server platforms to have multitenant or massive multitenant supports for cloud computing. On top of it is integration middleware, which is the foundation of EAI. The application-level integration middleware layer is also called SES by some research firms, and together with the platform middleware they become application frameworks for different vertical applications. Some packaged applications such as ERP, Supply Chain Management (SCM), Manufacturing Execution System (MES), and others are built on top of those middleware frameworks. The multitiered WoT architecture can also use the PaaS/SaaS technology. This will be described in the following chapters.

To summarize, many individual standards development organizations want to incorporate existing standards into a unified conceptual framework as much as possible following the same approach described in this chapter. Rather than reinvent what already exists, these organizations prefer to identify and fill gaps and to integrate what already exists into the unified horizontal framework described in Chapter 3. This approach recognizes that it is impossible, or at least undesirable, to try to define new physical layer technologies, networking layer protocols, or platform middleware-based application frameworks for every current or future potential WoT/M2M application. Different vertical applications will optimize for individual cost and functionality requirements, while a standardized service layer will facilitate cross-vertical application development.

ABI Research believes that initial proposals for such a unified framework and service layer could be available by early 2012. It would likely take another 18–24 months for this initial proposal to be formally published as a standard or set of standards. ABI Research doesn't expect these efforts to start having an impact until late 2013 to early 2014. When such standards are in place, they will play an important role in driving overall WoT/IoT market development.

7.3 WoT Portals and Business Intelligence

A web portal or links page is a website that functions as a point of access to information in the World Wide Web. A portal presents information from diverse sources in a unified way. At the beginning of the web-based Internet revolution, web portals played a crucial role in making the web popular among the general public. Examples of public web portals include Yahoo, AOL, Excite, MSN, and more recently, iGoogle. Apart from the standard search engine feature, web portals offer other services such as e-mail, news, stock prices, information, databases, and entertainment.

In the portal craze of the late 1990s, the web portal was a hot commodity. After the proliferation of web browsers in the late 1990s, many companies tried to build or acquire a portal to have a piece of the Internet market. Netscape became a part of America Online, the Walt Disney Company launched Go.com, IBM and others launched Prodigy, Excite and @Home became a part of AT&T, and so forth.

There are two broad categorizations of portals: horizontal portals, which cover many areas, and vertical portals, which are focused on one functional area. A vertical portal called vortal consequently is a specialized entry point to a specific market or industry niche, subject area, or interest. WoT portals are vertical portals.

By the same token, WoT portals also started to appear; some of the well-known ones are as follows:

- Pachube (https://pachube.com): Pachube ("patch-bay") (renamed Cosm recently), the "Plumber" of Internet, connects people to devices, applications, and the Internet of Things. As a web-based service built to manage the world's real-time data (has been used to monitor the radiation in Japan caused by the 2011 earthquake and tsunami), gives people the power to share, collaborate, and make use of information generated from the world around them.

- SensorMap (Microsoft, http://atom.research.microsoft.com/sensewebv3/sensormap/): The portal and its accompanying tools will allow for more online live data. Microsoft Hohm is another WoT project similar to Google PowerMeter, which is to be discontinued in 2012.
- Google PowerMeter (http://www.google.com/powermeter/about/): PowerMeter, retired in September 2011, included key features like visualizations of energy usage, the ability to share information with others, and personalized recommendations to save energy.
- Sun SPOT (small programmable object technology): Programming the world with Java, the Oracle Sun SPOT project explores wireless transducer technologies that enable the emerging network of things, building a hardware and software research platform to overcome the challenges that currently inhibit development of tiny sensing devices.

As intranets grew in size and complexity, webmasters were faced with increasing content and user management challenges. A consolidated view of company information was judged insufficient. Users wanted personalization and customization. *EIPs (enterprise information portals)* also became common after the public portals. EIP solutions can also include workflow management, collaboration between work groups, and policy-managed content publication. Most can allow internal and external access to specific corporate information using secure authentication or single sign-on.

Java Specification Request (JSR168) standards emerged around 2001. JSR168 standards allow the interoperability of *portlets* across different portal platforms. These standards allow portal developers, administrators, and consumers to integrate standards-based portals and portlets across a variety of vendor solutions. The concept of content aggregation seems to continue gaining momentum, and portal solutions will likely continue to evolve significantly over the next few years. The Gartner Group predicts generation 8 portals to expand on the business mashups concept of delivering a variety of information, tools, applications, and access points through a single mechanism. This technology should also be considered in WoT applications; for example, the ^{ez}M2M middleware platform developed by the author used the JSR168 standard portlet technology, mostly for Intranet IoT applications. As an example, an IoT platform with EIP portal and dashboard support can be found at http://iobridge.com/.

On the other hand, there is a need for a set of *ontologies* to marry sensor data and sensing information with meaning. The W3C's working group on semantic sensor networks is currently developing some definitive examples using RDF metadata model and related technologies discussed in the previous chapter. Additionally, real-time extension of the semantic sensor web concept is being developed, called Sensor Wiki. The motivation behind this concept is to allow real-time browsing of the physical world consistent with the situational awareness goal. Understanding the physical world via a myriad of sensors is now possible.

In a sensor Wiki, one or more sensors contribute real-time information as Wiki pages with suitable themes and formats useful to prospective Sensor Wiki users. Sensor Wiki users can look up information about objects, events, or places of interest interactively. They can also add intelligent interpretations of what they observe, use sensor tasking to add to the content to improve accuracy, or even develop the overall scene to offer situation assessment on a proactive basis. Others might want to record such sensor streams and related information as part of a larger objective such as future planning, training, or simply record keeping for historical purposes, and make it available to a specific community or an individual.

On a more practical basis, when enormous amount of data are collected in a IoT system, data mining can be conducted to acquire business intelligence (BI) and help decision support. Data mining deals with finding patterns in data that are by user definition, interesting, and valid. It is an interdisciplinary area involving databases, machine learning, pattern recognition, statistics, visualization, and others. Decision support focuses on developing systems to help decision-makers solve problems. Decision support provides a selection of data analysis; simulation; visualization; modeling techniques; and software tools such as decision support systems, group decision support and mediation systems, expert systems, databases, and data warehouses.

BI technologies provide historical, current, and predictive views of business operations. Common functions of BI technologies are extract, transform, and load (http://ckbooks.com/computers/data-warehousing/extract-transform-load-etl/) as well as reporting, online analytical processing, analytics, data mining, process mining, complex event processing, business performance management, benchmarking, text mining, predictive analytics, and so on.

For example, in many SCADA applications, BI is widely used. SCADA software vendors provide a number of relevant products such as CitectSCADA Reports, Wonderware Intelligence, Acumence Plant Analytics Server, and others. Also, a BI analytics of data from a fleet management system in China's truck/bus industry (partner of the author's current company) reveals that fuel usage can differ as much as 15 percent due to driver behavior for the same truck, route, and mileage.

7.4 Challenges of IoT Information Security

The security issue of the IoT is always an issue of concern, just like the security issue of all ICT systems. In the context of the IoT, because most of the "Things" (devices, assets, equipment, facilities, etc.) are owned by certain entities, the *ownership* characteristics make the security concern of IoT systems even more significant, often more tricky to deal with than the existing documents processing dominant ICT systems. For example, the whereabouts and size of an RFID-tagged and tracked nuclear warhead on the move could be exposed on the Internet if the security system is hacked. Privacy, such as the location of an object or a person, is one of the most concerning issues. Also, the sheer number of devices to be managed will add complexity to the existing security measurements. The Internet of Things will no doubt present new security challenges in cryptographic security, credentialing, and identity management.

Some technologically disadvantaged countries are worried about the security issues more than others. In China, for example, experts are evaluating potential new security threats brought in by the broad implementation of the Internet of Things across national borders. They are worried about the connected IoT systems such as the power grid, transportation (railways, airways, and roads) systems, water supply system, oil and gas pipelines, and so forth being compromised by third parties and losing *information sovereignty*, considering that developed countries such as the United States have programs such as the U.S. Army Signal Command (ASC), which provides support for the war-fighting commanders to win the information war. ASC directs the activities of some 15,000 soldiers and civilians in more than a dozen nations around the world. The USANETCOM (USA Network Enterprise Technology Command) [135] makes the continental U.S.-centered army capable of executing a force-projection mission through its integrated, worldwide theater tactical information assets. In theaters outside the continental United States, the USANETCOM provides the total spectrum of information services through centralized operation and maintenance of European, Southwest Asian, Pacific, and Central American strategic, theater, tactical, and sustaining-base information systems. On the other hand, developed countries are accusing underdeveloped countries of hacking into their systems to steal information and technologies.

However, there is no fundamental difference between IoT security and the traditional ICT system security. People have been putting hard-earned money (the most important assets) in banks and access accounts and do transfers via the Internet without much problem. The security measurements of the current ICT systems have eight dimensions [237]: access control, authentication, nonrepudiation, data confidentiality, communication security, data integrity, availability, and privacy. These technologies still apply to IoT systems and cover most, if not all, of the IoT security concerns and requirements, especially at the early stages of IoT development.

Some of the specific security issues that are more concerned with IoT application scenarios include the following (as summarized by the Association for Automatic Identification and Mobility, taking the RFID scenario as an example but applicable to all IoT systems):

- Skimming: Data are read directly from the tag without the knowledge or acknowledgment of the tag or device holder.
- Eavesdropping or sniffing (also called "man-in-the-middle" reader): Unauthorized listening/intercepting.
- Data tampering: Unauthorized erasing of data to render the device useless, or changing the data.
- Spoofing: Duplicates device data and transmits it to a receiver to mimic a legitimate source.
- Cloning: Duplicates data of one device to another device.
- Malicious code: Insertion of an executable code/virus to corrupt the enterprise systems.
- Denial of access/service: Occurs when multiple devices or specially designed devices are used to overwhelm a receiver's capacity, rendering the system inoperative.

- Killing: Physical or electronic destruction of the device deprives downstream users of its data.
- Jamming: The use of an electronic device to disrupt the receiver's function.
- Shielding: The use of mechanical means to prevent reading of a tag or device.

Due to the above issues and the capacity of devices and diverse networking conditions, a few challenges face the development of IoT in addition to traditional ICT security issues, especially at the advanced stages of IoT development:

- The 10 security issues listed above and the sheer number of devices involved will make the design and deployments of security solutions more complex.
- The heterogeneous, multihop, distributed networking environments make the passing and translation of security credentials and the end-to-end security functionalities a very difficult mission across the four categories of networks, that is, the long- and short-range wireless, and the long and short wired networks categorized in the previous chapters of this book.
- These cryptographic suites were designed with the expectation that significant resources (e.g., processor speed and memory) would be available. The differences of sizes, limited storage capacities, and constrained processing power of the devices also make the processing of public key infrastructure (PKI) encryption, decryption, and key management hard to be consistent along the entire data flow.
- The joining and leaving (bootstrapping) of devices into the IoT systems and the grouping of the mobile devices over dynamic networks also add complexity to the authentication and authorization process.

Some of the following security protocols are discussed as candidate solutions in the 6LoWPAN and CoRE IETF working groups [136].

- The Internet Key Exchange (IKEv2)/IPsec and MOBIKE (Mobility and Multi-homing IKEv2)
- The Host Identity Protocol (HIP) and a HIP variant for lossy low-power networks called Diet HIP
- Transport layer security (TLS) and its datagram-oriented variant DTLS secure transport-layer connections
- The Extensible Authentication Protocol (EAP)
- The Protocol for Carrying Authentication for Network Access (PANA)

Secure Middleware for Embedded Peer-to-Peer systems (SMEPP) is an EU/ICT project that built a security middleware framework for IoT applications [238].

KoolSpan's TrustChip® (http://www.koolspan.com/) is a fully hardened, self-contained security engine that aims to provide an end-to-end security solution over resource-constrained, heterogeneous networks.

7.5 Summary

Compared with data format standardization, the standardization of a unified IoT system architecture is more feasible and doable, especially the back end multitiered platform middleware architecture. This is one of the important conclusions drawn in this chapter.

Many projects worldwide but mostly in Europe have created a number of architectural specifications for IoT that cover one or more of the four pillar segments, some with reference implementation prototypes. Some companies such as

Tridium worldwide but mostly in the United States have also announced platform middleware products such as the Niagara Framework and ArchestrA of Wonderware for generic IoT applications. The aforementioned conclusion was drawn based on the investigation and analysis of those efforts and cases.

This chapter pointed out that the unified IoT architecture should be based on the existing multitiered middleware architecture, especially the JavaEE three-tiered application server architecture with support of related technologies such as SOA, ESB/EAI, OSGi, and others without reinventing the wheel. Other relevant technologies such as BI, information security, data formats, and more were also discussed in this chapter.

We will be talking about cloud computing and its synergy with IoT in the next two chapters.

THE CLOUD OF THINGS

Chapter 8

Cloud Computing

8.1 What Is Cloud Computing?

"It starts with the premise that the data services and architecture should be on servers. We call it cloud computing—they should be in a 'cloud' somewhere. And that if you have the right kind of browser or the right kind of access, it doesn't matter whether you have a PC or a Mac or a mobile phone or a BlackBerry or what have you—or new devices still to be developed—you can get access to the cloud." This is the vision of Google chief executive officer Eric Schmidt, speaking at a search engine conference in 2006. Since then, *cloud computing* has been a buzzword worldwide [277]. This was the first high-profile usage of the term; however, the first mention of cloud computing was in a 1997 paper entitled "Intermediaries in Cloud-Computing: A New Computing Paradigm" by R. Chellappa [276].

The term *cloud* was used as a metaphor for the Internet, based on the cloud drawing used in the past to represent the telephone network, and later to depict the Internet in computer network diagrams as an abstraction of the underlying infrastructure it represents.

Much like the Internet of Things (IoT), cloud computing is a natural evolution of related, existing, and new concepts in the information and communications technologies (ICT) arena, based on the widespread adoption of virtualization (first paper published in 1959 by C. Strachey [138]), cluster computing [139,140,141], grid computing, service-oriented architecture (proposed by Gartner in 1996), web services, parallel and distributed file systems [150], load balance and batch scheduling [142], autonomic, and utility computing technologies. In fact, the cloud computing term collided with many other terms that were already catchphrases in the ICT industry, such as SaaS (software as a service), grid computing, utility computing, PaaS (platform as a service), on-demand services, pervasive computing, and so on. Cloud computing provides computation, software, data access, and storage services that do not require end-user knowledge of the physical location and configuration of the system that delivers the services. Details are abstracted from end-users, who no longer have the need for expertise in, or control over, the technology infrastructure "in the cloud" that supports them.

The underlying concept of cloud computing dates back to 1961, when John McCarthy proposed the concept of utility computing and IBM started to rent its mainframe computing resources as "electric utility" to Wall Street via remotely connected dumb terminals.

Amazon played a key role in the development of cloud computing. After the dot-com bubble, most data centers were using as little as 10 percent of their capacity at any one time, just to leave room for occasional spikes. Having found that the new Amazon infrastructure resulted in significant internal efficiency improvements, Jeff Bezos initiated a new product development effort to provide cloud computing to external customers, and launched Amazon Web Service on a utility computing basis in 2006, which is now categorized as IaaS (infrastructure as a service).

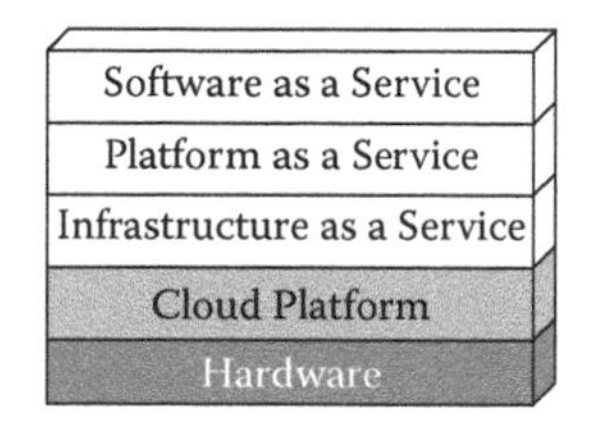

Figure 8.1 Cloud hierarchy. (From Ralf Teckelmann, Anthony Sulistio, and Christoph Reich, "A Taxonomy of Interoperability for IaaS," in Lizhe Wang, Rajiv Ranjan, Jinjun Chen, and Boualem Benatallah (Eds.), *Cloud Computing: Methodology, Systems, and Applications,* Boca Raton, FL: CRC Press, 2011.)

Salesforce was founded in 1999 by Marc Benioff, who is regarded as the leader of what he has termed the "no software" movement. In 2001, salesforce.com pioneered the multitenant SaaS model, a new application-delivering mechanism, a step beyond the application service provider model that made companies such as Exodus a great success in dot-com times. SaaS provides immediate benefits at reduced risks and costs, thanks to the rapid development and maturity of the Internet infrastructure, among other factors (as shown in Figure 8.1). In 2008, the force.com PaaS on-demand development platform was launched and became a new pillar of cloud computing, which has three pillars: IaaS, PaaS, and SaaS. All three pillars can provide services.

8.2 Grid/SOA and Cloud Computing

Much like the Internet of Things, the technological foundation of cloud computing is distributed computing based on communication networks. One of the most directly related works is the use of cluster of workstations (COWs) and networks of workstations.

Google's server farms were based on the same philosophy of COW. In fact, the MapReduce system is a batch processing extension of the scatter–barrier–reduce primitives of MPI/PVM

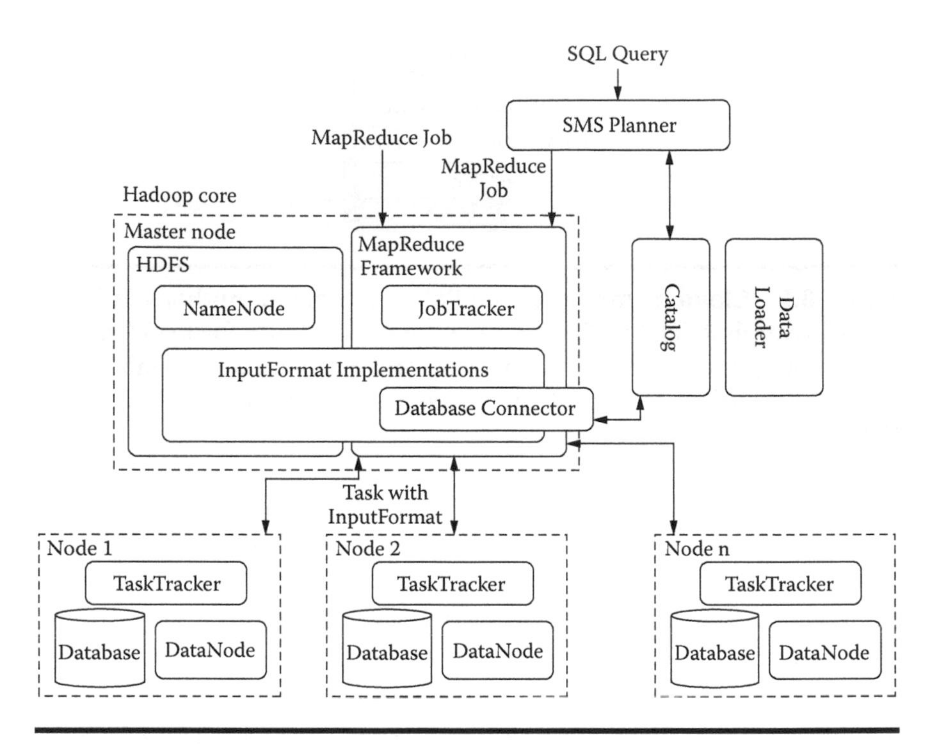

Figure 8.2 Hadoop is a batch system for embarrassingly parallel niche applications.

[143,144]. The well-known Hadoop is an open-source implementation of Google's Bigtable, GFS, and MapReduce [145,146,147] by Doug Cutting et al. in 2004. Hadoop is a high-throughput computing batch processing system, a niche application customized for embarrassingly parallel Internet-based massive data processing, as shown in Figure 8.2. However, as the Internet-related data and users increase rapidly, Hadoop has been widely used and become almost a nickname of cloud computing.

Grid computing is the direct technological ancestor of cloud computing, which also has roots in the COW technology; some of the well-known cloud systems such as Eucalyptus and OpenNebula are directly transformed from earlier grid computing research and development systems. (Much like *cloud*, the term *grid* is chosen as an analogy to the electrical power grid that provides consistent, pervasive, dependable,

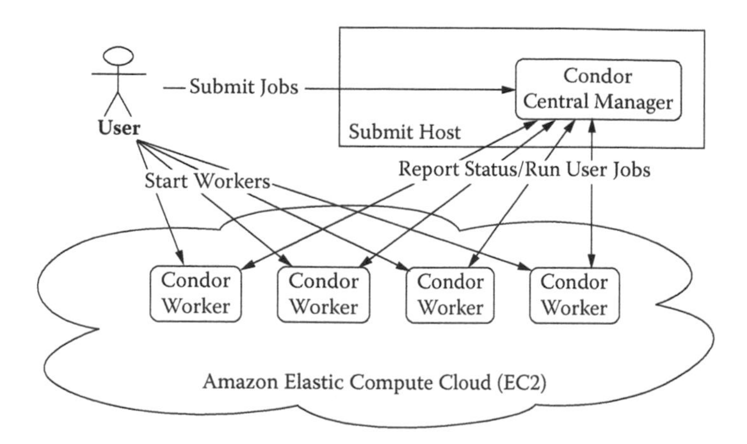

Figure 8.3 Condor scheduler in Amazon Web Services.

transparent access to electric power, irrespective of its source.) The grid computing concept was credited to Ian Foster of Argonne National Laboratory when he initiated the Globus project in 1994 based on the works of PVM/MPI, PBS/Condor [148,149] (as job schedulers of high-performance computing or parallel supercomputing systems), and so on. All of those technologies are generally referred to as cluster computing (other examples include Beowulf, Linux Virtual Server, MOSIX, BONIC) in a nutshell. Besides the basic parallel and distributed computing environments provided by middleware such as PVM and MPI, the job scheduler plays an important role as workload and resource management systems in building the grid and cloud computing/clustering infrastructure. For example, the Condor scheduler can be used in the Amazon system as shown in Figure 8.3. (The author worked in the LoadLeveler team, which was a job scheduler based on Condor [142], and participated in the ASCI–Blue Pacific project to build the then-world's-fastest massively parallel processing supercomputer, as the job scheduling system coordinator in 1996.)

One of the key features or functionalities of grid and cloud computing is providing a single system image (or a single parallel virtual machine) that hides the underlying scalable,

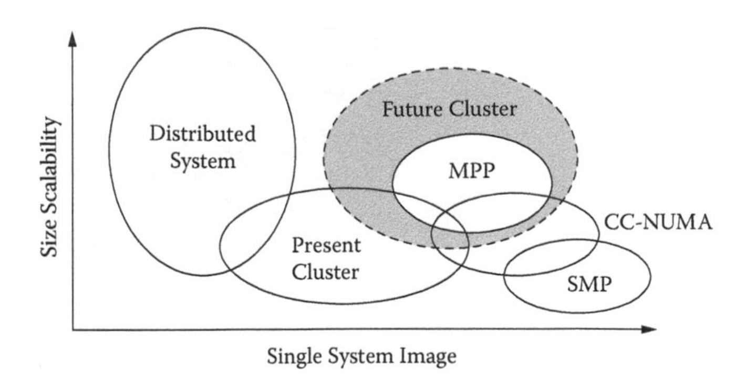

Figure 8.4 Single system image is the core.

elastic infrastructure (such as the Amazon backend server and storage farms) based on clustering technologies (as shown in Figure 8.4) and provisions a unified user interface via web services (such as the Amazon Web Services) and SOA over the Internet.

Virtualization is another important concept often mentioned in the cloud computing context. There are two sides of the virtualization coin: single system virtualization (SSV, i.e., one-to-many virtualization) and multisystem virtualization (MSV, i.e., many-to-one virtualization). This categorization of SSV and MSV for cloud computing was proposed by the author [75]. Utility computing started with SSV when IBM provided computing resources for rent via networked dumb terminals to Wall Street in the 1970s as mentioned before. A mainframe computer was virtualized into multiple virtual computers (as shown in Figure 8.5) via *hypervisor* technology so that

App1	App2	App3	App4
CMS	CMS	CMS	CMS
VM/370			
370 Hardware			

Figure 8.5 Earliest hypervisor.

enterprise users feel like a dedicated computer is providing services to them.

Modern SSV technologies are similar to the hypervisor technologies that IBM used decades ago. The purpose is to simulate multiple computers on top of one computer—run multiple operating systems on one computer hardware to enable maximum usage of the ever-increasing power of a single computer such as a PC and increase efficiency of overall resources. For example, one of the most important uses of SSV in earlier times was to simulate all the operating systems on a few servers so that a startup company in Silicon Valley could test their software products on all operating systems without having to buy all kinds of computers. SSV can be further categorized as three types of virtual machine monitors: Type 1 (hypervisor), Type 2, and hybrid [240].

The virtualization (currently almost a synonym of VMWare) talked about in the context of cloud computing currently is SSV, which makes many believe that SSV is a must for cloud computing. In fact, one-to-many virtualization is not required to build a cloud computing system, although it enables new platforms to run on legacy environments, and it helps to consolidate and simplify the management of the system by making the nodes of the system homogenous, thus simplifying the handling of issues such as (fault tolerance) check-pointing and migration (e.g., VMWare vMotion) when some of the nodes run into failures.

On the other hand, many-to-one virtualization is the foundation of cloud computing. MSV refers aggregately to the use of the aforementioned distributed and parallel clustering technologies such as COW, high-performance computing (HPC), grid computing, high-throughput computing, high-availability computing, and so forth to build a single, gigantic, parallel virtual computer or a single centralized service-providing virtual resource that serves many users for a plethora of applications. Some sample MSV architectures are shown in Figure 8.6. SSV

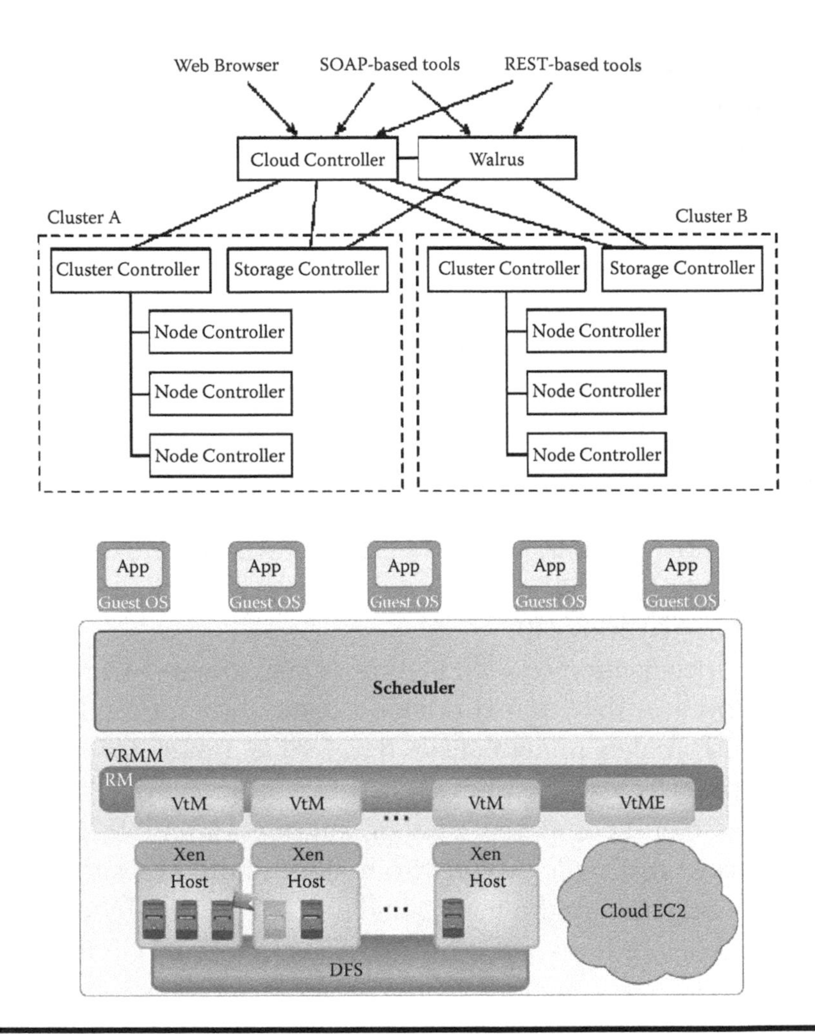

Figure 8.6 Many-to-one virtualization is the foundation of cloud computing.

technologies can be used at the node level of MSV but are not required.

The computing and storage resource are delivered to the end users using SOA (including SOAP or REST-based web services, SaaS, EAI, etc.) via the Internet, sometimes via intranet and extranet for private cloud applications. Many of the technologies and protocols of the SOA standard stack [241] can be used in all of the three layers: IaaS, PaaS, and SaaS.

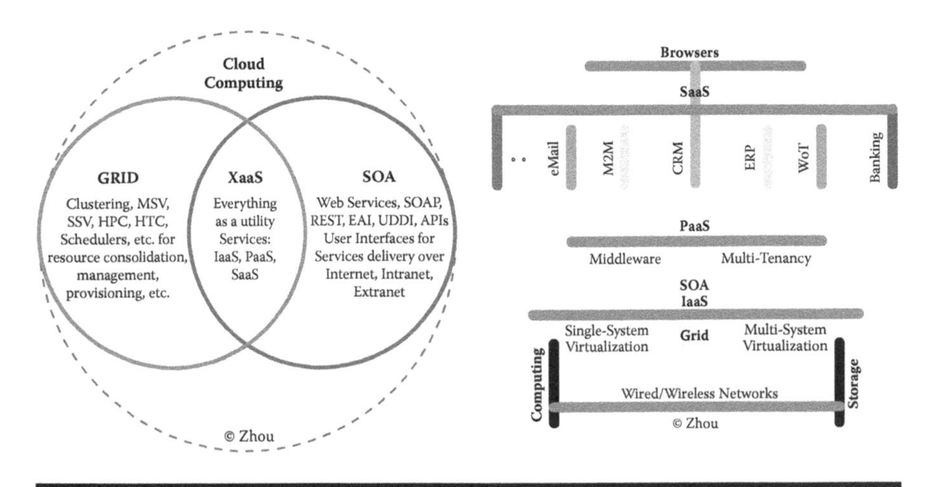

Figure 8.7 Cloud computing is the fusion of grid and SOA.

To summarize, cloud computing is the fusion of grid computing and SOA technologies to provide everything as utility-style services, as shown in Figure 8.7 [75]. It is "a large-scale distributed computing paradigm that is driven by economies of scale, in which a pool of abstracted, virtualized, dynamically-scalable, managed computing power, storage, platforms, and services are delivered on demand to external customers over the Internet," as defined by Ian Foster [278]. The graphic on the right of Figure 8.7 is based on the Chinese word for *cloud* that depicts the technologies, resources used, and application models.

8.3 Cloud Middleware

Again, much like the Internet of Things, the cloud computing system is also a multitiered architecture built on a middleware stack as shown in Figure 8.8.

At the lowest machine virtualization (SSV) level, there are middleware that help reduce the overhead of virtualization. SSV is useful and widely used, but it does not come cheap. The performance cost of virtualization, for I/O-intensive workloads in particular, can be heavy. Common approaches to

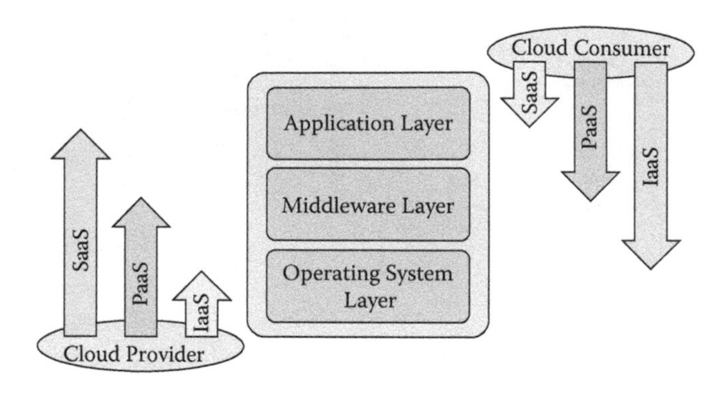

Figure 8.8 Multitiered cloud architecture based on middleware.

solve the I/O virtualization overhead focus on the I/O stack, thereby missing optimization opportunities in the overall stack. As an example, VAMOS [242], built by IBM, is a novel middleware architecture that runs its middleware modules at the hypervisor level. VAMOS reduces I/O virtualization overhead by cutting down on the overall number of guest/hypervisor switches for I/O intensive workloads. Applying VAMOS to a database application improved its performance by up to 32 percent. Here, the middleware concept is extended to include software that does interprocess communication not necessary over a network.

At the cluster computing or grid computing level, many types of work are done by middleware. The parallel computing environments such as PVM and MPI are (HPC) middleware by definition; the Hadoop system and the job scheduler such as Condor, LoadLeveler, and others are all middleware. The HPC middleware fills the gap that the operating systems and the programming languages lack to support parallel computing [151]. A number of grid middleware initiatives (such as http://www.eu-emi.eu/) have been formed by interested members, mostly in the scientific computing community. Some of those middleware are aggregately referred to as grid middleware [152,153] and listed as follows:

- Low-level middleware
- MPI, Open MPI
- PVM (parallel virtual machine)
- POE (parallel operating environment, IBM)
- Middleware for file systems and resources
- MPI-IP
- PVFS/GPFS (parallel virtual file system/general parallel file system IBM)
- Sector-Sphere
- Condor/PBS/LoadLeveler (IBM)
- High-level middleware
- Beowolf
- Globus Toolkit
- Gridbus
- Legion
- Unicore
- OSCAR/CAOS/Rocks
- OpenMosix/NSA/Perceus

Many research works [243] demonstrate a typical grid computing system (other similar systems include Distributed European Infrastructure for Supercomputing Application [DEISA], Teragrid, Enabling Grids for E-Science [EGEE], NorduGrid, SEE-GRID, OSG, etc.) and its components based on grid middleware before cloud computing gained momentum. A grid computing system aims to serve all kinds of applications as a more generic cloud computing system than Hadoop.

As discussed before, grid computing is the foundation of cloud computing infrastructure, so grid middleware is the basis of IaaS middleware. In addition, the IaaS middleware (part of cloud middleware [244]) may include components such as system management, network management, billing and operation support systems, provision, configuration, automation, orchestration, service level agreement (SLA) management, and so on.

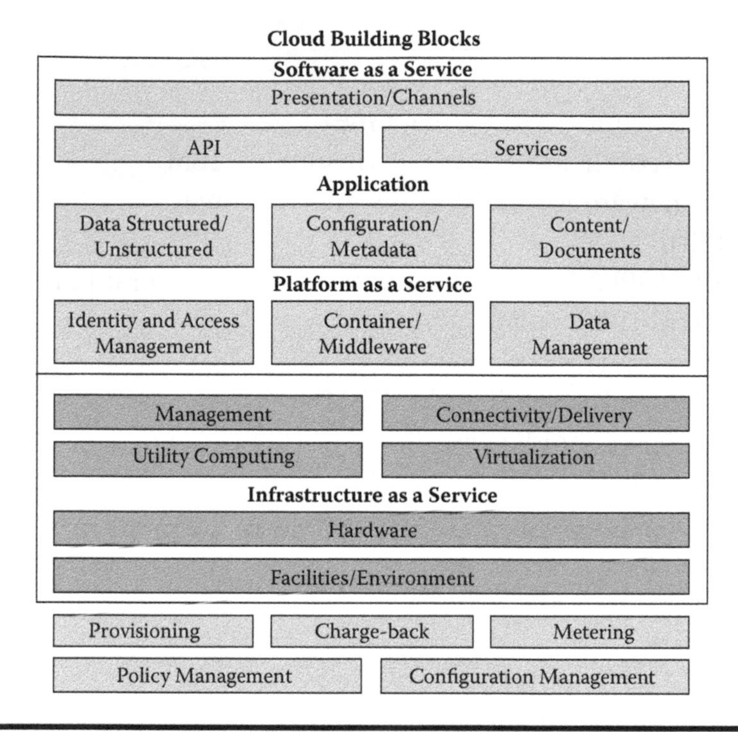

Figure 8.9 PaaS and cloud middleware.

From the distributed enterprise computing standpoint, almost all of the EAI and business-to-business (B2B) middleware described in the previous chapters are needed to build cloud computing systems for enterprise and commercial applications. They all are part of cloud middleware, particularly part of the PaaS middleware. Multitenancy [245] is one of the basic functions of PaaS middleware, evolving from the traditional platform middleware. The multitenant efficiency functionalities of a PaaS platform are often required and implemented in a traditional middleware such as the three-tiered application servers described in the previous chapters and as shown in Figure 8.9. More comprehensive guides on all of the building blocks of cloud computing have been discussed and depicted [267,268].

The PaaS middleware is often referred to as the cloud middleware that underpins and supports the SaaS applications.

The graphic in [246] depicts the different deployment options of a PaaS middleware in cloud systems: the more middleware is shared, the cloud systems scale to larger numbers of tenants and with lower operational costs.

The well-known middleware [247] quadrant from Gartner depicts the market landscape of middleware vendors. Salesforce.com was included for the first time in 2010, most likely because its foundational platform (force.com) is recognized as one of the most important cloud (PaaS) middleware vendors.

To summarize, the cloud middleware consists of two kinds of middleware—IaaS and PaaS middleware—and their relation is shown in Figure 8.10. (Note: SaaS are not middleware, they are applications on top of middleware.)

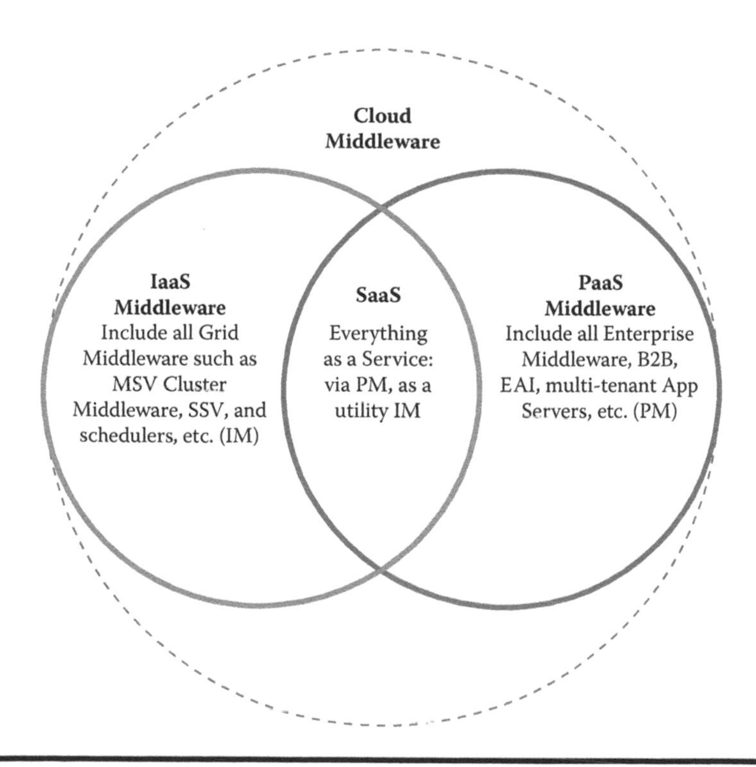

Figure 8.10 Cloud middleware.

8.4 NIST's SPI Architecture and Cloud Standards

The U.S. National Institute of Standards and Technology (NIST) has come up with a widely accepted definition [154] that characterizes important aspects of cloud computing: Cloud computing is a model for enabling ubiquitous, convenient, on-demand network access to a shared pool of configurable computing resources (e.g., networks, servers, storage, applications, and services) that can be rapidly provisioned and released with minimal management effort or service provider interaction.

This cloud model is composed of the following:

- Three service models: IaaS, PaaS, and SaaS
- Four deployment models: private cloud, public cloud, community cloud, and hybrid cloud
- Five essential characteristics: on-demand self-service, broad network access, resource pooling, rapid elasticity, and measured service

An earlier version (Version 14) of the specification also listed 12 foundational elements or enablers.

Cloud computing is an evolving paradigm. The comparative benefits of the different service models of cloud computing are compared in http://itcandor.net/2010/11/22/cloud-computing-benefits-q410/. The NIST specification is a milestone that clarifies and settles most of the confusion and arguments about cloud computing. It can be used as a starting point for standardization. Electronics and Telecommunications Research Institute (ETRI) of Korea proposed to address standards on nine aspects (http://www.etri.re.kr/eng/res/res_0102020301.etri):

- Definition, taxonomy, terminologies
- Provisioning model

- Business process
- Security
- Interoperability
- Legality
- Environmental issues
- Architecture
- Availability

The NIST specification covers a few of the aspects, such as the standardization of definition, taxonomy, and terminologies.

Some of the standardization in the grid computing domain provided a foundation for extended work, such as the MPI, openMP standards, as well as job description language standards (such as Job Submission and Description Language [155] and Basic Execution Service [156] of Open Grid Forum—Open Grid Services Architecture) for job scheduling.

Table 8.1 lists some of the cloud computing standardization organizations and their websites. The following are some of the works done by those standards organizations:

- NIST: Working definition of cloud computing
- Distributed Management Task Force: Open Virtualization Format, Open Cloud Standards Incubator, DSP-IS0101 Cloud Interoperability White Paper V1.0.0
- Cloud Management Working Group: DSP-IS0102 Architecture for Managing Clouds White Paper V1.0.0, and DSP-IS0103 Use Cases and Interactions for Managing Clouds White Paper V1.0.0
- European Telecommunications Standards Institute: TC cloud definition
- Standards Acceleration to Jumpstart Adoption of Cloud Computing: 25 use cases
- Open Cloud Consortium: Open Cloud Testbed, Open Science Data Cloud, benchmarks, reference implementation

Table 8.1 List of Standardization Efforts

List of Standardization Efforts
National Institute of Standards and Technology (http://csrc.nist.gov/groups/SNS/cloud-computing/index.cfm)
Distributed Management Task Force (http://www.dmtf.org)
The European Telecommunications Standards Institute (http://www.etsi.org)
Open Grid Forum (http://www.ogf.org)
Open Cloud Computing Interface Working Group (http://www.occi-wg.org)
Object Management Group (http://www.omg.org)
Storage Networking Industry Association (http://www.snia.org)
Open Cloud Consortium (http://www.opencloudconsortium.org)
Organization for the Advancement of Structured Information Standards (http://www.oasis-open.org)
Association for Retail Technology Standards (http://www.nrf-arts.org)
The Open Group (http://www.opengroup.org)
Cloud Security Alliance (http://www.cloudsecurityalliance.org)

- The Cloud Computing Interoperability Forum: framework/ontology, semantic web/resource description framework, unified cloud interface
- The Open Group: SOA, The Open Group Architecture Framework
- Association for Retail Technology Standards: Cloud Computing White Paper V1.0
- TM Forum: Cloud Services Initiative, Enterprise Cloud Leadership Council Goals, Future Collaborative Programs, BSS/OSS/SLA
- ITU-T FG Cloud: Introduction to the Cloud Ecosystem: Definitions, Taxonomies and Use Cases;
- Global Inter-Cloud Technology Forum: Japan, Interoperability

- Cloud Standards Coordination: Standards Development Organization Collaboration on Networked Resources Management
- Open Cloud Manifesto (http://www.opencloudmanifesto.org/)
- Open Grid Forum: Open Cloud Computing Interface, Open Grid Services Architecture
- Cloud Security Alliance: Security Guidance for Critical Areas of Focus in Cloud Computing, Cloud Controls Matrix, Top Threats to Cloud Computing, CloudAudit
- Storage Networking Industry Association: Cloud Storage Technical Work Group, Cloud Data Management Interface

8.5 Cloud Providers and Systems

In five short years, cloud computing has gone from being a quaint technology to being a major catchphrase. It started in 2006 when Amazon began offering its Simple Storage Service and soon following up with its Elastic Compute service, and Google's CEO Eric Schmidt's speech about cloud computing. Just like the Internet of Things, the market potential is huge. Many vendors, old and new, have joined the gold rush to provide cloud services and products. There are many forecasts about market size of cloud computing. For example, Gartner estimated that, among the three SPI segments, SaaS generates most of the revenue, because it directly creates value for the end users. IaaS helps reduce the costs of organizational users, which has the fastest growth. Gartner predicts the change of revenue on percentage among the three SPI segments between 2010 (SaaS: 72%, PaaS: 26%, IaaS: 2%) and 2014 (SaaS: 61%, PaaS: 36%, IaaS: 3%). However, this prediction may not count the revenue of PaaS as a middleware product sold independently, but only the part of the revenue of PaaS as a hosted service, in which case PaaS is sold as part of SaaS most of the times.

Revenue generated by cloud technology companies, excluding the larger, more mature SaaS segment, is forecast to grow from $984 million in 2010 to $4 billion in 2013, according to The 451 Group, representing a compound annual growth rate of 60 percent. Including SaaS, total cloud technology vendor revenue was $8.5 billion in 2010, expected to grow to $16.3 billion in 2013, a compound annual growth rate of 24 percent. Of course, the amount spent by companies on cloud products and services is much larger, with Gartner estimating worldwide cloud services revenue in 2010 of $68.3 billion, an increase of 16.6 percent from $58.6 billion of 2009. Gartner estimates that the cloud services revenue will reach $148.8 billion in 2014.

We will give an overview of the current cloud providers based on their participation in providing the building blocks as depicted in the graphics noted below that include the services and products for the three SPI pillars, as well as additional products such as development tools, security frameworks, system management software, adaptor frameworks, and so on.

There are many top 10, top 20, and top 50 listings of cloud providers on the web that can be easily found with Google search. The graphic at http://www.opencrowd.com/assets/images/views/views_cloud-tax-lrg.png [157] lists some of the top cloud providers in the SPI and general software categories.

Gartner published its Magic Quadrant (http://www.cloudbusinessreview.com/wp-content/uploads/2011/01/magic-quadrant-gartner.png) about the leading IaaS providers and the emphasis (http://blogs.pcmag.com/miller/assets_c/2010/10/Cloud%20Vendor%20Emphais-16535.php) of the best-known cloud providers. Another well-known cloud vendor taxonomy graphic is from Peter Laird of Oracle (BEA) created in 2009, which can be found at http://farm4.static.flickr.com/3312/3597138202_496ae06a68_o.png.

The cloud computing boom has brought a surge of opportunity to the open-source world. Open-source developers and users are taking advantage of these opportunities. Many open-source applications are now available on a SaaS basis. Other open-source projects have taken the steps necessary to make them easy to use in the cloud, for example, by making preconfigured images available through Amazon Web Services or other public clouds. However, most open-source developers are contributing to the growth of cloud computing by creating the tools that make cloud computing feasible. They offer infrastructure, middleware, and other software that make it easier for companies to develop and run their applications in the cloud. The following is a list of open-source projects:

- Open-source IaaS and PaaS projects: OpenStack, cloud.com Cloud Stack, OpenNebula, Eucalyptus, AppScale, Scalr, Traffic Server, RedHat Cloud, Cloudera (Hadoop), Puppet, Enomaly, Joyent, Globus Nimbus, Reservoir, Amanda/Zmanda, XCP, TPlatform, and so forth
- Open-source SaaS projects: Zoho, Phreebooks, Pentaho, Palo BI Suite, Jaspersoft, Processmaker, eyeOS, Alfresco, SugarCRM, SourceTap, KnowledgeTree, OpenKM, Collabtive, Zimbra, Feng Office, Open ERP, Openbravo, Compiere, Orange HRM, JStock, Ubuntu, OpenProj, openSIS, TimeTrex, GlobaSight, and others

To summarize, the author has created a free-style panoramic view graphic [75] of existing cloud providers and their products and services in five layers (including vendors and products in China that are mostly at the PaaS and SaaS layers):

- Chip and hardware supports for virtualization: Intel-VT (VT-x, VT-x2), AMD-V (SVM), SUN/Oracle UltraSPARC T1, T2, T2+, SPARC T3, and others
- Hypervisors (one-to-many SSV virtualization) vendors and products
- IaaS (many-to-one MSV virtualization) grid/cluster computing, web services–based delivery vendors and products

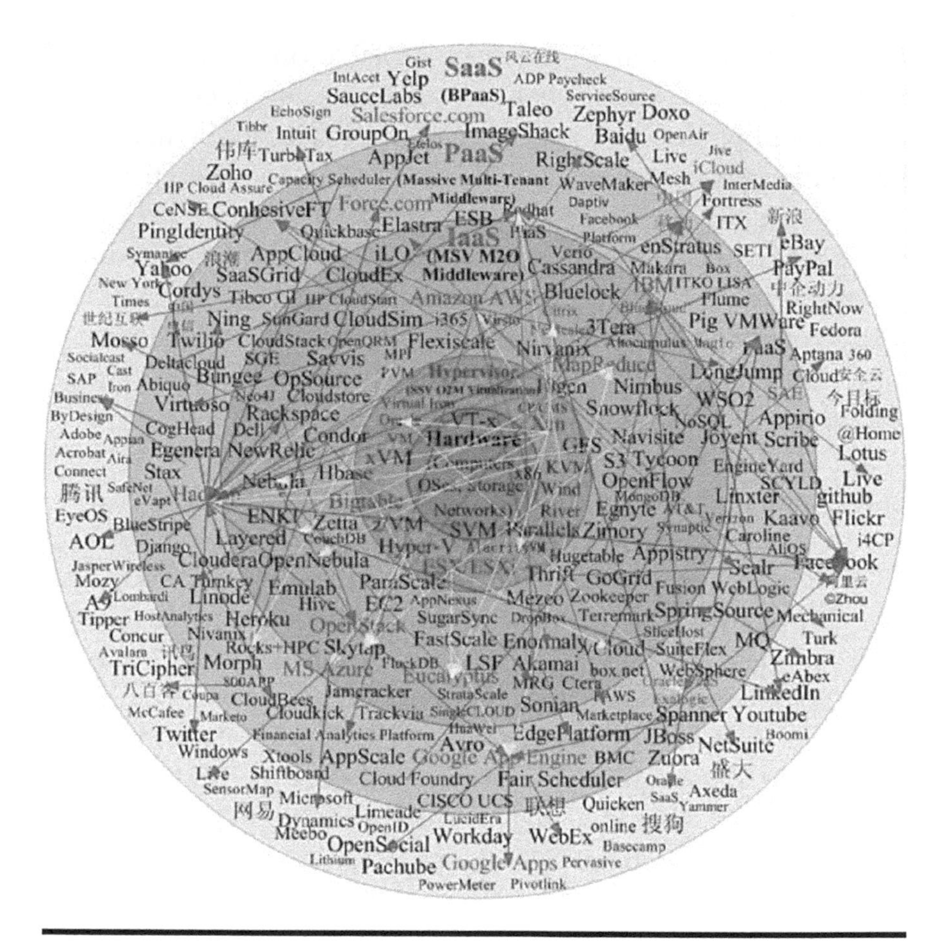

Figure 8.11 Five-layer panoramic view of cloud vendors and products.

- PaaS (multitiered middleware) vendors and products
- SaaS vendors and products (due to the vastly large number of SaaS vendors and products, only some of them are listed; some of the IoT SaaS services such as Pachube are also listed. See Figure 8.11.)

8.6 Summary

In this chapter, we talked about what cloud computing is, its relationship with earlier concepts, and paradigms such as grid computing, cluster computing, SOA, SaaS, and the like. The importance of middleware in cloud computing is described and emphasized. The systematic specification of NIST and many standardization efforts were introduced and discussed. And finally, a comprehensive summarization of the currently existing vendors, service providers, and systems is provided.

Much like cloud computing, the Internet of Things is also about distributed computing. The two have many things in common and many shared underlying technologies and paradigms, which will be discussed in the next chapter.

Chapter 9

The Cloud of Things

9.1 The Internet of Things and Cloud Computing

The Internet of Things (IoT) and cloud computing are two of the most widely used catchphrases nowadays in media. In the English-speaking world, however, the term *Internet of Things* is not as popular as *cloud computing*, as discussed before and also evidenced by the Google Trends chart (Figure 9.1). Part of the reason is that IoT is referred to by different terms such as machine-to-machine (M2M), connected world, smarter planet, smart grid, and the like in the United States.

However, Google trends (Figure 9.2) show that *machine to machine* is a more popular term than *cloud computing*, while *M2M* is less popular.

Whatever the situation is, both IoT and cloud computing can be categorized as distributed computing and have many things in common or closely related:

- Both are a type of distributed computing that relies heavily on communication networks.
- Cloud computing is an enabling technology of the IoT.

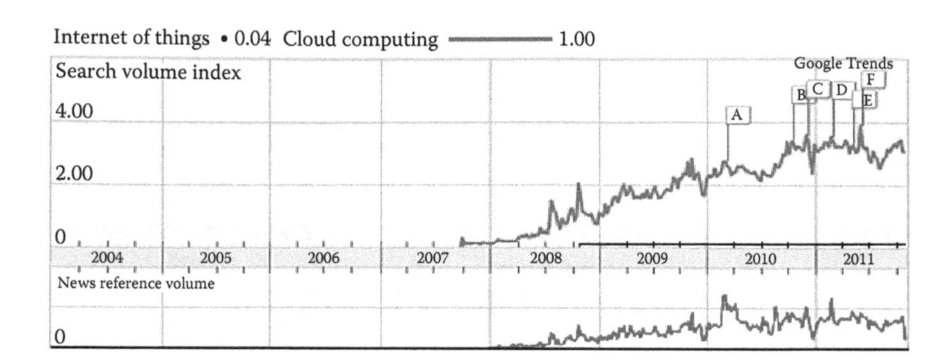

Figure 9.1 IoT versus cloud computing.

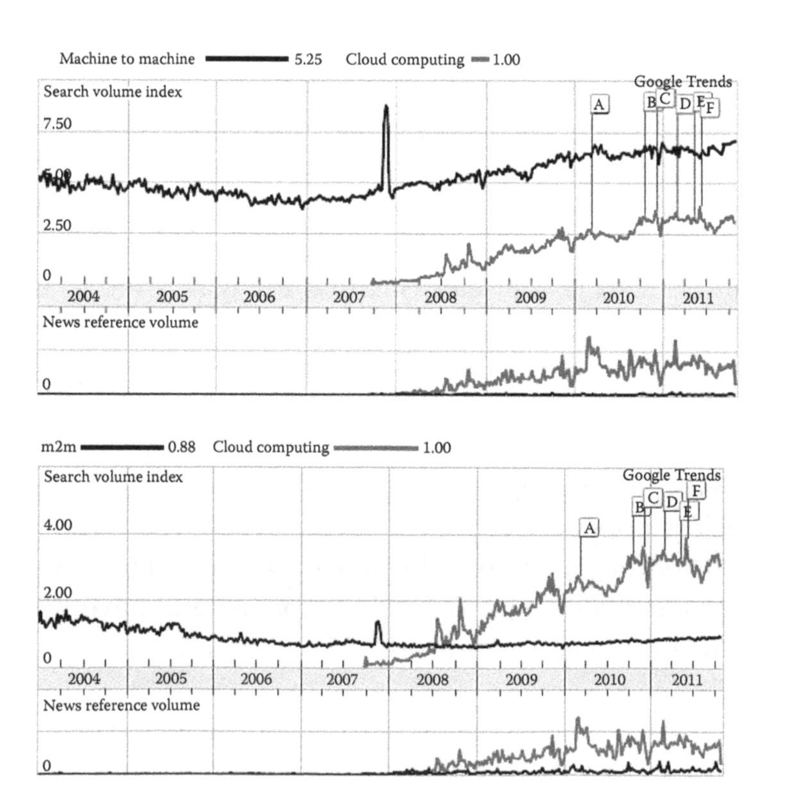

Figure 9.2 M2M versus cloud computing.

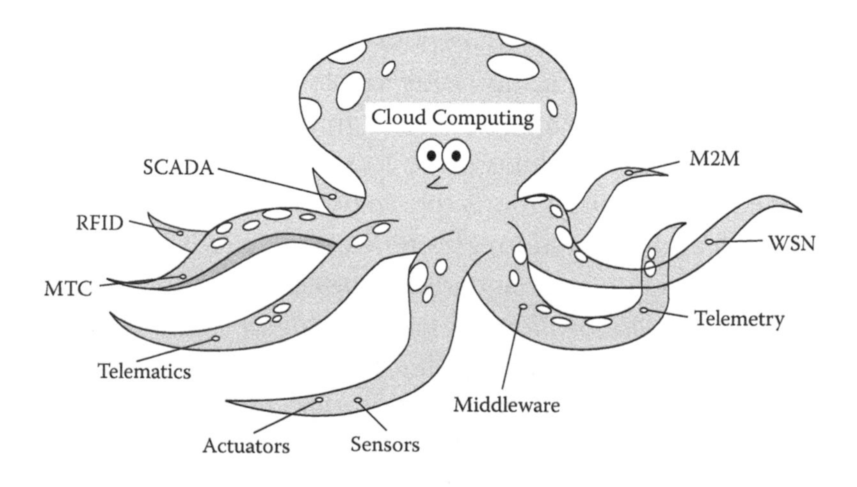

Figure 9.3 Cloud computing and IoT.

- The cloud and IoT are best considered as a continuum of Internet connectivity with cloud as (focusing on) the "head" and IoT as the "tails" of an octopus as shown in Figure 9.3.

We are in the early stages of the Internet of Things, the much-anticipated era when all devices can talk to intermediary services and to each other. But for this era to achieve its full potential, operators must fundamentally change the way they build and run clouds [158]. The reason is that M2M interactions are far less failure tolerant than machine-to-human interactions. Imagine when a fleet of trucks can no longer report its whereabouts to a central control system designed to regulate how long drivers can stay on the road without resting, or when the power in your building goes out and the heating, ventilation, and air-conditioning (HVAC) system dies on a hot day because of a cloud outage.

In the very near future, everything from elevators to cell phones to city buses will either be subject to connected control systems or use networks to report back critical information. The sheer volume of data flowing through networks will mushroom. In a dedicated or colocated hardware world, that would result in prohibitively expensive hardware requirements. Thus, the cloud becomes the only viable option to affordably connect, track, and manage the new Internet of Things.

Current M2M/IoT solutions are focusing on communications (i.e., how information is transmitted from one machine to another) and integration. Future Web of Things (WoT) evolution will effectively integrate *connectivity* and *content* with *context*, *collaboration*, *cloud*, and *cognition*. The future Internet of Things will be a global network of interconnected objects, enabling object identification/discovery and semantic data processing via the M2M-IOT C^6 cube depicted in Interdigital.com website [137], with cloud as the base.

- Connectivity: connection for mobile and constrained objects
- Content: massive data produced from things
- Cloud: cloud service and cloud content storage
- Context: context-aware design to improve performance
- Collaboration: cooperative communications, inter-things, service sharing
- Cognition: mine the knowledge from massive data and provide autonomous system adjustment for improvements

In this expanded role, the cloud will have to step up its game to accommodate more exacting demands. The current storage infrastructure and file systems that back up and form the backbone of the cloud are archaic, dating back 20 years. These systems are familiar and comfortable for infrastructure providers. But over time, block-storage architectures that cannot provide instant snapshots of machine images (copy-on-write) will continue to be prone to all sorts of failures. Those failures will grow more pronounced in the M2M world when a five-second failure could result in the loss of many millions of dollars worth of time-specific information.

Currently, no one is putting truly mission-critical applications in the cloud. But in the coming era of the Internet of Things, that is a near-guaranteed eventuality, either through intentional or unintentional actions. As we build the Internet of Things and slowly ease it first onto private clouds and later onto public clouds, we have no choice but to improve the core of the cloud or risk catastrophic consequences from failures. Because on the Internet of Things, no one can blame it on user error and simply ask that a hotel air conditioner, an airplane, or a bank of traffic lights restart its virtual server on the fly and reset its machine image.

In short, the Internet of Things will not take off without an up-to-date, secure, and scalable cloud computing infrastructure such as the ones from Eurotech Everyware (http://www.eurotech-inc.com/m2m.asp), Jelastic (http://blog.jelastic.com/2012/01/09/using-jelastic-for-the-Internet-of-things/), and so on.

9.2 Mobile Cloud Computing

The potential of cloud computing doesn't stop at turning the personal computer into a thin client. The mobile application market is about to change radically, from the suppliers' standpoint and from the consumer access standpoint due to the emergence of widgets (applications from Apple app stores or Android markets, a ranking is available at http://www.pocketberry.com/2011/02/18/blackberry-app-world-gets-2nd-place-for-global-mobile-app-store-ranking/), the most compelling of mobile cloud applications. Much has been made of the mobile application phenomenon popularized by Apple's iconic iPhone. Smartphones are becoming thin clients of cloud services, which render software and content vendors such as Microsoft, Google, and Apple into the upper streams of the smartphone value chain. Traditional cell phone makers such as Nokia, Sony-Ericsson, HTC, and others are lagging behind and struggling.

Apple's iCloud services, announced in June 2011 that run on Amazon Web Service and Microsoft Azure IaaS, symbolize the start of Cloud Phones (even though Apple's iTunes has been a cloud service for a long time), which is followed by traditional mobile phone firms and Internet services companies such as Google worldwide. With all or most of the computing and heavy-lifting done on the server side, cell phones become a device that handles connectivity. This will also bring down the price of smartphones. A cloud architecture for smartphones is envisioned by NTT DoCoMo [248].

Currently, most widgets downloaded from app stores or Android markets are not cloud applications by definition because they do not receive services from the cloud during runtime. However, a large number of them are cloud applications such as LBS applications, data synchronization, weather forecast, bank client, etc., applications. In fact, a large percentage of Android and iPhone widgets are already cloud services based. This is real mobile cloud computing (mCC). Apple's iCloud services allow users to store data such as music files on remote computer servers for download to multiple devices such as iPhones, iPods, iPads, and personal computers running Mac OS X or Microsoft Windows. Windows Live Mesh is a free-to-use Internet-based file synchronization application by Microsoft that is designed to allow files and folders between two or more computers to be in sync with each other on Windows and Mac OS X computers.

Android Cloud to Device Messaging (C2DM) [249] is a service that helps developers sending data from servers to their applications on Android devices. The service provides a simple, lightweight mechanism that servers can use to tell mobile applications to contact the server directly, to fetch updated application or user data. The C2DM service handles all aspects of queuing of messages and delivery to the target application running on the target device.

Mobile cloud widget applications such as AppStore widgets as we know them today are mostly for the domain of smartphone users. Gartner estimated the total sales of smartphones across 2011 were 472 million or 31 percent of mobile communication. The rest of the mobile subscriber world has generally had to stand by and watch, since their phones are not powerful or fast enough to handle mobile apps. Nevertheless, non-smartphones or so-called feature phones can still get connected with the cloud to receive simple services such as data synchronization, etc.

Smartphone apps are typically custom built for particular smartphone platforms in advanced programming languages, limiting the available pool of developers and driving up costs. Using a stand-alone widget client is in fact a step back technologically due to the limitations of the browsers on the smartphones in the earlier days. With HTML5, the widget client may again be unified with one browser client just as it happened on PCs while at the same time keeping the revenue generation model of charging the users based on application widget download.

The HTML5 has features such as offline support, canvas drawing based on low footprint SVG graphics, GeolocationAPI, video and audio streaming support without flash, WebStorage, CSS3 Selectors, 2D animations for mobile cloud applications. Widgets, whether in the forms of app stores or unified under a browser, will exponentially expand the market for mobile applications with the heavy-lifting done in and contents to and from the cloud (some call it mobility-as-a-service [250]), introducing complex, rich user experiences to a new and much larger mobile consumer audience.

Those mobile cloud computing services have made the "phone" a "thing" in the Internet and they can be easily extended for other IoT applications such as Google Wallet and mobile resource management (MRM, as described in Chapter 2). For example, some of the mobile devices such as telematics terminals are both a M2M device connected to the cloud and a smartphone thin client that receives services from the cloud. ADT's Pulse is a project associated with Android@Home. It not only allows you to arm and disarm your ADT security system, but also includes very impressive controls for lights, security cameras, and even thermostats. Another example is

Figure 9.4 Schlage IoT application.

the Schlage LiNK iPhone applications that can let you turn off your home lights remotely (Figure 9.4) while enjoying a vacation in Hawaii. The parent company of Schlage is Ingersoll Rand, which also owns Trane, a heating, ventilation and air-conditioning (HVAC) company. This relationship also allows the Schlage LiNK to work easily with a Trane thermostat, and so on.

It's fair to say that, without M2M and sensor capabilities, mobile cloud computing has limited value. MCC, especially the so-called micro cloud computing systems such as Hyrax, and sensor networking are in fact very similar technology paradigms.

As IP-enabled, affordable sensor devices of all types become available and are placed around the earth forming a "sensing cloud," integrating the diverse sensor data streams into the web can serve different user or machine queries. In the sensorMap project of Microsoft and the Pachube project, people are encouraged to contribute real-time sensor information to the cloud subject to privacy and security constraints. Intelligent mobile devices can act as hubs or sources and sinks of such real-time streams as shown in the Pachube ecosystem graphic [251].

It's not hard to imagine that, over time, all of the phones and mobile devices will become thin clients that receive cloud services and smart devices that send information such as location or environment data to the cloud, to finally achieve the full potentials of M2M; i.e., machine to machine, machine to mobile, machine to man connectivity.

According to ABI Research, 19 percent of global mobile users will be using cloud-enabled devices in 2014. Juniper Research forecast that the revenue of mobile cloud computing applications will increase from $400 million in 2009 to $9.5 billion in 2014. The predictions (new computing cycles support by 10X more devices) of Morgan Stanley can be found in "Internet Trends" [252].

With the development of software technologies and the increase of the processing power of the devices at lower cost, specific protocols such as WAP for mobile devices are no longer needed. The difference between mobile and non-mobile devices are narrowing, different devices are converging in usage and information delivery methods. For example, smartphones and the telematics terminals as well as the tablet PCs (iPad) could be one unit in the future, TVs, PCs, and smartphones can run the same software suite.

Telecom operators have been investing big money to build cloud infrastructure for M2M applications:

- Verizon Wireless and Sierra Wireless have announced a new collaboration to co-market Sierra Wireless' AirVantage (described in Chapter 4), a cloud-based platform for developing, deploying, and operating the next generation of connected devices and M2M applications;
- AT&T is working with Axeda to build cloud-based applications for telematics, security solutions, monitoring, supervisory control and data acquisition (SCADA), point of sale, asset management and similar M2M deployments. AT&T's innovative service delivery platforms complement its network and expertise for a broad range of wireless data applications and industries;
- In China, China Mobile (Big Cloud), China Telecom (Nebula), and China Unicom are all building cloud computing to support iCloud/iPhone-like "Cloud Phone" and M2M applications.

In the future, with the all-IP technologies such as LTE, cloud phones may become virtual personal phones hosted in the network with a IPv6 address that are accessed via a personal ID from any device without the need of a SIM card. Phone calls become a functionality of a smart M2M device. Cloud services are moving toward serving smaller smart devices with support of robust middleware platforms as forecasted by ABI Research.

For example, announced in December 2010, the Amazon Web Services SDK for Android provides a library, code samples, and documentation for developers to build connected mobile applications. Similar SDK also exists for iPhone, iPad, and iPod Touch devices.

The Open Mobile Terminal Platform (OMTP) was a forum created by mobile network operators to discuss standards with manufacturers of cell phones and other mobile devices. In July 2008, OMTP announced an initiative called Bondi (http://bondi.omtp.org/). The initiative defined new standard-based, vendor-agnostic interfaces (Javascript APIs) and a security framework (based on XACML policy description) to enable the access to mobile phone functionalities (application invocation, application settings, camera, communications log, gallery, location, messaging, persistent data, personal information, phone status, user interaction) from a browser and widget engine in a secure way. This effort would certainly help in building widespread mobile cloud computing (mCC) services.

GSM Association's third-party access OneAPI (http://oneapi.gsmworld.com/) is another standardization effort that aims to provide a set of open-network enabler APIs (OneAPI) that can be supported across mobile operators and other networks. OneAPI is based on lightweight RESTful and SOAP APIs to encourage portability of mobile apps but still allow for competition and differentiation between operators. As mentioned before, HTML5 also provides better support for mobile applications, which might render the widget-based approach currently used by AppStore and Android markets unnecessary.

Ubiquitous connectivity is creating great market opportunity and unlimited application potentials per an ABI Research forecast [253], with Asia-Pacific sharing the largest piece of the pie. Telco operators, equipment, and smart device (including cell phones) manufacturers are taking strategic steps to embrace the mCC and pervasive computing opportunities.

Mobile computing, cloud computing, and IoT are intertwined with each other, like the many facets of a diamond. The core is connectivity and software-enabled resource sharing and services.

9.3 MAI versus XaaS: The Long Tail and the Big Switch

As discussed in Chapters 6 and 7, the difference between EAI and business-to-business/business-to consumer (B2B/B2C) is that one is for internal Intranet and the other is for external Internet integration. The concepts of M2M application integration (MAI) and XaaS (Everything as a Service) were proposed [74]; they are the extensions of EAI and B2B/B2C respectively in the IoT space.

Today, the majority of IoT devices live in the MAI systems that exist in the Intranet and Extranet. Only a fraction of the devices are available on the Internet. The focus of MAI is connectivity and monitoring. In the future, XaaS of IoT will provide more services to a larger audience, as shown in Figure 9.5.

The Long Tail theory was popularized by Chris Anderson in an October 2004 *Wired* magazine article [119], in which he mentioned Amazon and Netflix as examples of businesses exploiting the Long Tail strategy and making enormous profits. The *Long Tail* refers to the statistical property that a larger share of population rests within the tail of a probability distribution than observed under a "normal" distribution. Anderson believes that IoT finally makes sense after so many years [257].

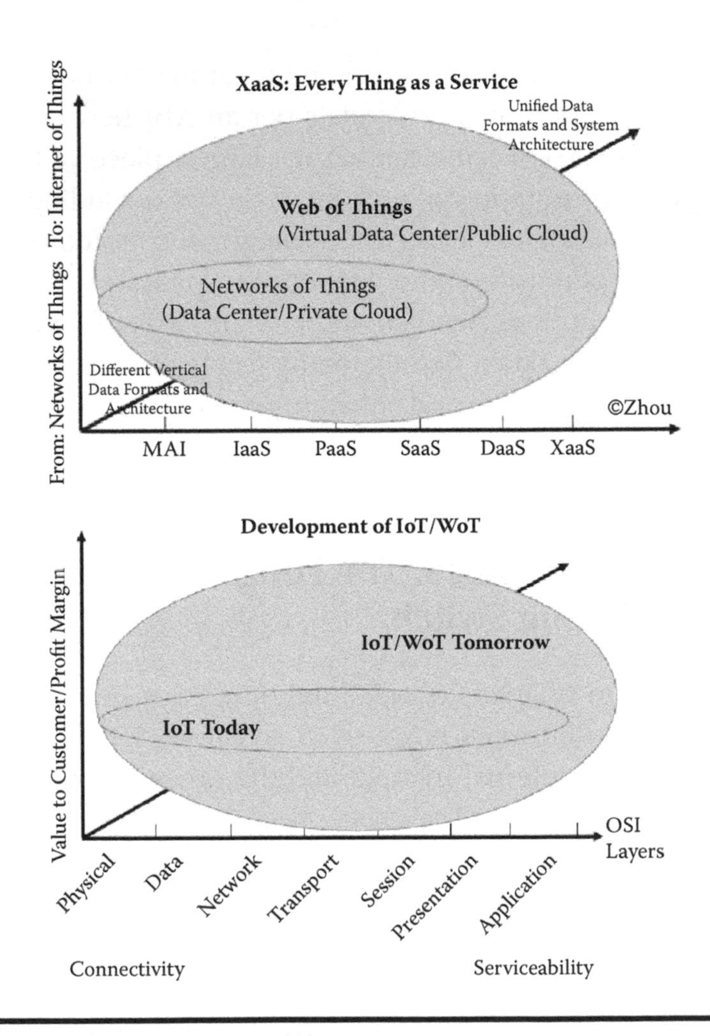

Figure 9.5 Evolution of IoT in the cloud.

The web technology on top of the Internet makes the harvesting of Amazon's online book-selling business Long Tail cost-effective. Based on Figure 9.5, the majority of currently connected *things* are located in organization's intranets, which form the Long Tail of the Internet of Things; only a minority exist in the Internet or extranet. Examples of *things* that are on the Internet are meters or sensors in Google's Powermeter, Microsoft's Sensormap, or on Pachube.com; the list is not long, and most of the projects are currently experimental.

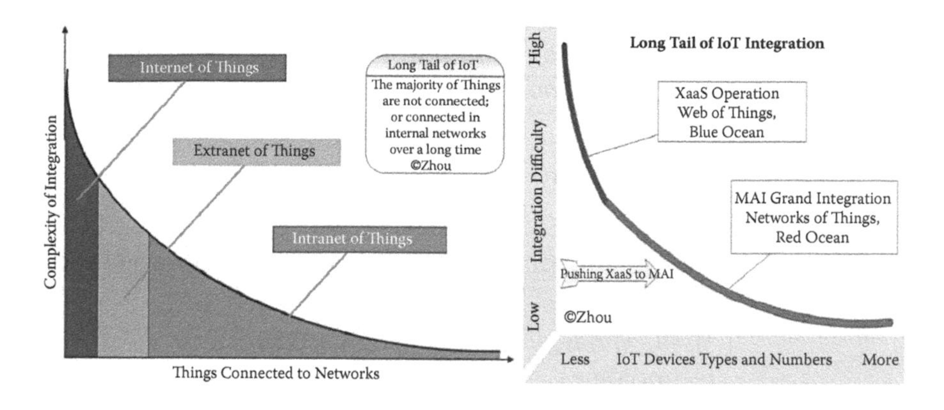

Figure 9.6 The Long Tail of IoT.

By the same token, with the constant development of the Internet of Things, more and more *things* or *objects* will be connected to the intranet, extranet, and finally the Internet with proper security measures, making the harvesting and utilization of the IoT Long Tail cost-effective and secure. (The author is one of the first who had this observation [74] and proposed the intranet, extranet, IoT concepts as depicted in Figure 9.6.) New innovative business models like that of Amazon and Netflix of today will emerge, and the ubiquitous IoT applications will become widespread and prosper.

At the beginning of IoT development, many people thought it was a Blue Ocean [168] opportunity just like when the web and the browser were invented. As we can see now, the MAI of IoT is an extended application of EAI, and the XaaS is an extension of web applications that cover devices and things. Most of the current IoT applications are foundational works that involve "Red Ocean" competitions. However, gold mining opportunities do exist when more and more ubiquitous devices are connected to the web; new application paradigms and new business models are on the horizon, as shown in Figure 9.7.

In *The Big Switch: Rewiring the World, from Edison to Google*, Nicholas Carr [279] walks readers through the history

IoT Strategy ©Zhou	*Red Ocean Competition The Foundation*	*Blue Ocean Strategy The Opportunity*
Device	Device and sensor design and manufacturing, cost reduction	Sensors based on new materials, R&D breatkthrough
Connect	Networking reliability, QoS 3G/4G and new technologies All-IP/IPv6 networks	3-Networks convergence, new models New paradigms: No SIM card, etc.
Manage	Middleware development MAI grand integration XaaS cloud services	New service model innovation Unified system framework Unified data formats

Figure 9.7 IoT Red Ocean versus Blue Ocean.

of electrification and computing. The early years of electrification were technologically limited. An electrical grid wasn't feasible and electricity was generated locally. Technology changed over time and electricity was rapidly centralized and networked. Power was produced remotely and delivered via a vast network of wires and cables.

Based on this historical context, Carr revitalized the metaphor between electrification and the current model of computing that was introduced by John McCarthy in 1961. We have come all the way to the time of cloud computing; computing resources can be used and charged just as electricity is consumed and billed. Examples are Amazon's EC2 (Elastic Computing Cloud) and S3 (Simple Storage Services). To people in the computing industry, Carr's sayings aren't new but thought provoking. In fact, the transportation networks also changed the consumption of goods fundamentally: food isn't homemade, vegetables aren't home grown, and so forth.

With the Internet of Things comes the big paradigm switch: as described before, the power grid becomes a two-way electricity supply, a smart grid system where people can store their electricity surplus generated by their solar system back to the smart grid. The ubiquitous Internet of Things makes the consumption of everything possible, just like electricity and computing resources. Examples include changing the driving route home in a telematics-enabled car by checking on a traffic-congestion map generated using sensor-based vehicle-to-road, vehicle-to-vehicle ITS systems, a service provided by a TSP, and so on. Thomas Friedman's "flat world" [169] will become more flattened and smarter with the Internet of Things [256] as a major flattening factor.

9.4 The Cloud of Things Architecture

Much like cloud computing, no agreed common terminologies, definitions, and architecture specifications for the Cloud of Things existed until Peter Mell and Tim Grace of the Information Technology Laboratory of the National Institute of Standards and Technology (NIST) proposed the NIST definition of cloud computing. Even though the European Union has created definitions, specifications, roadmaps, and so on, it seems that still no agreed definitions and architecture specifications exist.

As mentioned before, IoT and cloud computing have many comparable characteristics. For example, cloud computing has three layers: IaaS, PaaS, and SaaS (SPI). IoT also consists of three layers: devices, connect, and manage (DCM) or devices, networks, and applications (DNA). Cloud computing has public cloud, private cloud, hybrid cloud, and so forth. The IoT also has Intranet of Things, Extranet of Things, Internet of Things, and so on.

Even though the IoT concept and paradigm is still evolving, the basic pieces of the IoT puzzle have been generally agreed.

Mimicking the NIST specification of cloud computing, a tentative IoT architecture/framework specification is proposed in this chapter, hoping to help form a common reference architecture framework and common terminology. One of the foundations of this specification is the four-pillar categorization of IoT.

Figure 9.8 is the general framework of the Internet of Things. Its definition, attributes, characteristics, use cases,

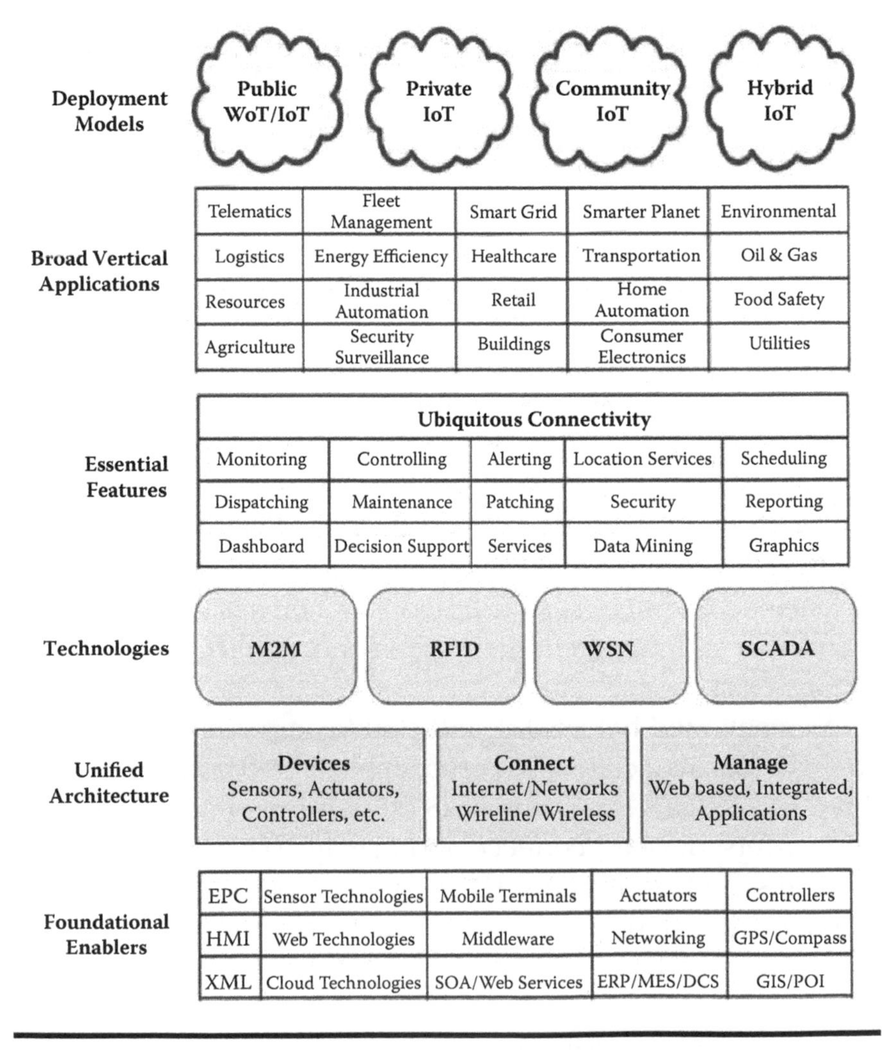

Figure 9.8 The Cloud of Things architectural specification.

underlying technologies, issues, risks, and benefits will be refined and changed over time in spirited debates by the public and private sectors. The IoT industry represents a large ecosystem of many models, vendors, and market niches. This specification attempts to encompass all of the various IoT approaches.

The definition of IoT by the author:

> The Internet of Things provides the means to access and control two categories of ubiquitous and uniquely identifiable devices—those that have inherent intelligence and those that are externally enabled—via all sorts of wired and/or wireless communications in all kinds of networking environments, supported by cloud computing technologies with adequate security measures, to achieve pervasive connectivity and grand integration and to provide services such as monitoring, locating, controlling, reporting, decision support, and so on.

9.4.1 Four Deployment Models

- Private IoT: The IoT MAI system is operated solely for an organization such as a building management system operated by a property management firm. It may be managed by the organization or a third party and may exist on premise (intranet) or off premise (extranet).
- Public IoT: The IoT system is made available to the general public or a large industry group and is owned by an organization, such as Pachube, selling IoT services.
- Community IoT: The integrated system is shared by several organizations and supports a specific community that has shared concerns (e.g., mission, security requirements, policy, and compliance considerations). It may be managed by the organizations or a third party and may exist on premise or off premise.

- Hybrid IoT: The IoT system is an integrated composition of two or more of the above IoT systems (private, community, or public) that remain unique entities but are bound together by standardized or proprietary technology that enables data and application portability.

9.4.2 Vertical Applications

Because there are too many vertical applications, it is impossible to list all of them. Only the vertical applications that are expected to be materialized soon are listed. For example, in China, telematics is expected to receive policy support from the central government sooner than the others. According to MIIT (Ministry of Industry and Information Technology) of China, during the 12th Five-Year Plan period, MIIT will spend more effort to promote telematics in full swing. It is said that telematics has been listed as one of three major projects supported by the central government (State Council) and will receive special financial funding. Informed sources said that support funding will focus on automotive electronics, telematics, fleet management (e.g., mostly due to security reasons, all heavy trucks and long-distance buses nationwide were mandated to be tracked and monitored by the Ministry of Transportation by the end of 2011), ITS, and so forth; and more than 10 billion renminbi (RMB) yuan will be allocated by the central government to support this effort. It's estimated that there will be 200 million vehicles in China by 2020, and all of them (passenger and commercial vehicles) will be mandated to be connected by that time.

- Telematics, fleet management, transportation
- Smart grid, energy efficiency
- Smarter planet
- Environmental protection
- Logistics, retail
- Healthcare

- Security/surveillance
- Resources (such as water resource management, etc.)
- Industrial automation
- Home automation, buildings
- Food safety, agriculture
- Security surveillance
- Consumer electronics
- Utilities, oil, and gas

9.4.3 Fifteen Essential Features

The fundamental feature of IoT is *ubiquitous connectivity*. Other concrete features or functionality (some of them are merged in the list) include the following:

- *Monitoring* and *Controlling*: These are some of the fundamental functionalities of IoT applications, more monitoring than controlling.
- *Location Services*: Based on GPS/compass (Beidou) or other locating technologies such as RTLS.
- *Alerting*: Event-based alerting, sometimes triggering rule-based engine for actions.
- *Scheduling* and *Dispatching*: Time- and event-based scheduling and dispatching.
- *Maintenance* and *Patching*: Maintenance supports remote monitoring, refill, patching (software upgrade), and so forth. For example, equipment from all Kodak shops around the world can be connected in a DRM (device relation management) system.
- *Security*: Security framework is required to support access control, privacy, and so forth.
- *Reporting, Dashboard*: Reporting, trending, and dashboard are used for better management and decision making.
- *Data Mining, Decision Support*: Analysis of collected device data based on Business Intelligence (BI) algorithms and data mining for decision support.

- *Graphics*: Graphic display of dynamic data, work flows, equipment status, and so forth of real-world things.
- *Services*: All kinds of services, such as postsale services of equipment, vehicles, leasing support and controls, and others.

9.4.4 Four Technological Pillars

1. RFID: IoT starts with radio-frequency identification. It's more of an enabling technology that turns dumb things into traceable items via instrumentation. It can also be used as identification means for counterfeiting and other applications. The usage is unlimited.
2. Wireless Sensor Network (WSN): The last-mile nerves of IoT including OSN, BSN, and others. Information can be gathered at the M2M gateway for uplink integration. Some WSN systems can be stand-alone.
3. M2M: This is an area the telcos [184] are focusing on. Mobile terminals can be connected and integrated for MRM, telematics, fleet management, and other applications. When all networks become IP-based such as LTE, cell phones can be part of multifunctional smart devices that no longer require a SIM or other card.
4. SCADA: It includes IT-controls converged smart system and others, an escalation of control systems. It can be used in buildings, industrial automation, smart grids, and more.

9.4.5 Three Layers of IoT Systems

1. Devices: include ubiquitous intelligent devices (M2M terminals, WSN sensors, SCADA actuators, etc.) and dumb assets that can be RFID instrumented to be electronically traceable
2. Connect: include wired and wireless, long-distance and short-range telecommunication means
3. Manage: integrated applications that are based on middleware and cloud computing back end

9.4.6 Foundational Technological Enablers

1. EPC: all coding and identification technologies such as EPC, UID, UUID, and others
2. Sensor technologies: all kinds of sensors, large and small, that generate MTC data
3. Mobile terminals: all kinds of mobile devices that communicate via telco networks
4. Actuators and controllers: PLC, RTU, DCS, and others that connect via field buses
5. HMI: human–machine interface technologies include graphical control panels, PC-based panels, and so forth
6. Web technologies: include browser/HTML5 technologies on all kinds of devices
7. Cloud technologies: SPI-based backend technologies, multitenancy
8. SOA/web services: for B2B/B2C grand integration over the Internet as well as intranet and extranet
9. XML: provides universal data representation means
10. Middleware: all kinds of middleware for unified IoT framework
11. Networking: provides ubiquitous connectivity
12. GPS/compass: provides location services
13. GIS/POI: help provide location and navigation services
14. ERP/MES: receivers and users of IoT data, part of IoT grand integration

9.5 Summary

In this final chapter of the book, the synergy of IoT and cloud computing was discussed. Mobile computing, cloud computing, and IoT are intertwined with each other, like the many facets of a diamond. Mobile cloud computing pushes the convergence a step further. In the future, with all-IP technologies such as LTE, cell phones may become part of any smart

M2M devices with an IP/IPv6 address without the need of a SIM card.

Most IoT technologies and applications are not new; what's novice is ideologies brought about by IoT. Two new paradigms, MAI and XaaS, are introduced by the author to describe IoT (inside the firewall) and WoT (outside the firewall) systems.

When the deployment of IoT applications and the number of connected devices reach a critical mass and scale, fundamental, innovative, and disruptive changes will emerge, just like prosperity of the web has brought about the Internet revolution. The thought-provoking ideas of the Big Switch and the Long Tail theory have been cited to stimulate creative imagination inspired by the Internet of Things and cloud computing.

As a final summarization of the entire book, the Cloud of Things architectural specification was introduced and explained with the hope of creating a common vocabulary for the IoT community.

References

1. Erico Guizzo, "The Rise of the Machines," 2008, http://spectrum.ieee.org/robotics/industrial-robots/the-rise-of-the-machines/1.
2. Brian Bremner, "Service Robots: Rise of the Machines (Again)," 2011, http://www.businessweek.com/magazine/content/11_11/b4219032532458.htm.
3. "The Rise of the Machines," http://www.energysavingtrust.org.uk/Publications2/Corporate/Research-and-insights/The-rise-of-the-machines-a-review-of-energy-using-products-in-the-home-from-the-1970s-to-today
4. "Beyond SCADA: Networked Embedded Control for Cyber Physical Systems," http://www.truststc.org/scada/.
5. *Vision and Challenges for Realizing the Internet of Things*, European Union, 2010, ISBN 9789279150883.
6. "Extracting Value From the Massively Connected World of 2015," www.gartner.com/DisplayDocument?id=476440.
7. "Internet 3.0: The Internet of Things." Analysys Mason Limited, 2010.
8. Ovidiu Vermesan et al., "Internet of Things Strategic Research Roadmap," CERP-IoT, http://www.internet-of-things-research.eu/pdf/IoT_Cluster_Strategic_Research_Agenda_2011.pdf, 2011.
9. Kevin Ashton, "That 'Internet of Things' Thing," *RFID Journal*, 22, July 2009.
10. P. Magrassi et al., *Computers to Acquire Control of the Physical World*, Gartner research report T-14-0301, September 28, 2001.
11. Commission of the European Communities, "Internet of Things: An Action Plan for Europe," June 2009.
12. Casaleggio Associati, "The Evolution of Internct of Things," 2011.

13. "ITU Internet Reports: The Internet of Things—Executive Summary," 2005, http://www.itu.int/osg/spu/publications/internetofthings/.
14. "A Smarter Planet: The Next Leadership Agenda," http://www.ibm.com/ibm/ideasfromibm/us/smartplanet/20081106/sjp_speech.shtml, 2008.
15. "Government 2.0: The Smarter Planet Initiative and Obama's Inauguration Speech," http://aaronkim.wordpress.com/2009/01/21/government-20-the-smarter-planet-initiative-and-obamas-inauguration-speech/, 2009.
16. "Obama Announces $3.4 Billion in Grants for Smart Grid," http://liveearth.org/en/liveearthblog/obama-announces-billions-for-smart-power-grid, 2009.
17. Gerald Santucci, "The Internet of Things: A Window to Our Future," http://www.theinternetofthings.eu/content/g%C3%A9rald-santucci-internet-things-window-our-future, 2011.
18. B. Schilit, N. Adams, and R. Want, "Context-Aware Computing Applications" IEEE Workshop on Mobile Computing Systems and Applications, 1994.
19. Adam Greenfield, *Everyware: The Dawning Age of Ubiquitous Computing*, New Riders Publishing, 2006.
20. "Obama Says IT Is Critical to Transforming Healthcare," http://www.healthcareitnews.com/news/obama-says-it-critical-transforming-healthcare, 2009.
21. Hakan Soderstrom, "U-Korea, U-Japan, U-Fever," http://www.soderstrom.se/?p=24, 2008.
22. EPoSS, "Internet of Things in 2020: Roadmap for the Future," http://www.smart-systems-integration.org/public, 2008.
23. E. Brezis, P. Krugman, and D. Tsiddon, "Leapfrogging in International Competition: A Theory of Cycles in National Technological Leadership," *The American Economic Review*, 1993.
24. "CASAGRAS and The Internet of Things: Definition and Vision Statement Agreed," http://www.rfidglobal.eu/userfiles/documents/CASAGRAS26022009.pdf, 2009.
25. "SAP: Internet of Things: An Integral Part of the Future Internet," http://services.future-internet.eu/images/1/16/A4_Things_Haller.pdf, 2009.
26. CERP-IoT, "Internet of Things: Strategic Research Roadmap," http://www.grifs-project.eu/data/File/CERP-IoT%20SRA_IoT_v11.pdf, 2009.

27. Bruce Sterling, *Shaping Things*, MIT Press, 2005.
28. "High Confidence Software and Systems: Cyber-Physical Systems," http://blackforest.stanford.edu/eventsemantics/Gill-CPSWeek-WEBS.pdf.
29. Network and Information Technology Research and Development (NITRD), http://www.cra.org/govaffairs/blog/tag/nitrd.
30. "Clicks & Mortar: Web 4.0, The Internet of Things," Hammer Smith Group Research Report, http://thehammersmithgroup.com/images/reports/web4.pdf, 2009.
31. "Three Key Enablers for Broadband Wireless," http://www.telecom-cloud.net/2010/07/12/3-key-enablers-for-broadband-wireless, 2010.
32. "Machine-To-Machine (M2M) and Smart Systems Forecast, 2010–2014," Harbor Research, 2010.
33. M2M Research, http://www.beechamresearch.com/.
34. http://www.m2mexpo.com/.
35. Machina Research, http://www.machinaresearch.com/.
36. "The Internet of Things," https://www.mckinseyquarterly.com/The_Internet_of_Things_2538, 2010.
37. "Pervasive Internet and Smart Services Market Forecast," http://www.harborresearch.com/HarborContent/2009%20PIMF%20Brochure_2009.pdf, 2009.
38. "M2M/Embedded Market Overview, Healthcare Focus, and Strategic Options," http://www.telco2research.com/articles/EB_M2M-Embedded-Overview-Healthcare-Strategic-Options_Summary.
39. "Connected World 100, 2102," http://www.connectedworldmag.com/M2MTop100.aspx.
40. "Overview of Mobile Resource Management Systems (MRM) Market," http://events.eft.com/truckit/presentations/1ClemDriscoll.pdf.
41. "Automotive Industry Trends," http://www.altera.com/end-markets/auto/industry/aut-industry.html, 2010.
42. "Remote Product Services Extend Benefits of Machine-to-Machine Solutions," http://www.arcweb.com/research/strategy-reports/2011/08/remote-product-services-extend-benefits-of-machine-to-machine-solutions.aspx.
43. Michael Schagrin, U.S. Department of Transportation, "National VII Architecture: Data Perspective," www.its.dot.gov/press/ppt/2008TRB682_National%20Architecture.ppt, 2008.

44. T. Oda and K. Takeuchi, "Driving Safety Support System in UTMS 21," http://www.utms.or.jp/english/inter/paper/seoul06.pdf.
45. P. Carter et al., "Delivering Next-Generation Citizen Services," IDC Report, http://www.cisco.com/web/strategy/docs/scc/whitepaper_cisco_scc_idc.pdf.
46. "Complex Interactive Networks/Systems Initiative: Final Summary Report," http://www.azouk.com/212870/Complex-Interactive-NetworksSystems-Initiative-Final-Summa/.
47. U.S. Department of Energy, "Grid 2030: A National Vision for Electricity's Second 100 Years," http://www.ferc.gov/eventcalendar/files/20050608125055-grid-2030.pdf.
48. Jerry Li, "From Strong to Smart: The Chinese Smart Grid and Its Relation with the Globe," http://www.aepfm.org/link.php, 2009.
49. Sam Lucero, "Horizontal Standards for M2M," http://www.abiresearch.com/research_blog/1650, 2011.
50. "The Six Pillars," http://www.constructech.com/news/articles/article.aspx?article_id=5625.
51. "The EPCglobal Architecture Framework," http://www.gs1.org/gsmp/kc/epcglobal/architecture/architecture_1_4-framework-20101215.pdf, 2010.
52. Sam Lucero, "Maximizing Mobile Operator Opportunities in M2M," ABI Research, 2010.
53. "Ubiquitous Sensor Networks (USN)," ITU-T Report, http://www.itu.int/dms_pub/itu-t/oth/23/01/T23010000040001PDFE.pdf, 2008.
54. S. Soro et al., "A Survey of Visual Sensor Networks," http://www.hindawi.com/journals/am/2009/640386/, 2009.
55. A. Seema et al., "Towards Efficient Wireless Video Sensor Networks: A Survey of Existing Node Architectures and Proposal for A Flexi-WVSNP Design," http://mre.faculty.asu.edu/WVSNPsurvey.pdf, 2011.
56. "Body Sensor Networks: The Next Generation of Health Care," http://bsn2009.org/, 2009.
57. M. Wang et al., "Middleware for Wireless Sensor Networks: A Survey," http://www.ccf.org.cn/web/resource/8301.pdf, 2008.
58. A. Thiagarajan et al., "VTrack: Accurate, Energy-Aware Road Traffic Delay Estimation Using Mobile Phones," http://citeseerx.ist.psu.edu/viewdoc/download?doi=10.1.1.161.8484&rep=rep1&type=pdf, 2009.

59. T. Hartman, "The Convergence of Building Controls, IT," http://hpac.com/bas-controls/convergence-building-controls-0509/index.html, 2009.
60. "Global SCADA and Machine-to-Machine (M2M) via Satellite Markets," http://www.giiresearch.com/report/ns87493-global-scada.html, 2009.
61. C. Amarawardhana et al., "Case Study of WSN as a Replacement for SCADA, "http://ieeexplore.ieee.org/xpl/freeabs_all.jsp?arnumber=5429891, 2009.
62. S. Methley et al., "Wireless Sensor Networks, Final Report," http://stakeholders.ofcom.org.uk/binaries/research/technology-research/wsn3.pdf, 2008.
63. "Intelligent Nuclear Power IOT Solutions," http://www.datang-telecom.com/templates/08Solutions%20Content%20Page/index.aspx?nodeid=147&page=ContentPage&contentid=242, 2011.
64. Bob Emmerson, "Networks in 2015: A Vision and a Strategy," http://www.tmcnet.com/voip/0808/networks-in-2015-a-vision-and-a-strategy.htm, 2008.
65. "'Internet Kill' Switch and IPv9," http://3g4g.blogspot.com/2010/06/internet-kill-switch-and-ipv9.html, 2010.
66. Toon Norp, "Mobile Network Improvements for Machine Type Communications," http://docbox.etsi.org/Workshop/2010/201010_M2MWORKSHOP/06_M2MGlobalCollaboration/Norp_TNO_mobileNtwImprovements.pdf, 2010.
67. "State of the Satellite Industry Report," http://www.sia.org/PDF/2011%20State%20of%20Satellite%20Industry%20Report%20%28June%202011%29.pdf, 2011.
68. Machine 2 Machine, "Innovation in M2M," http://machine2twomachine.wordpress.com/2011/08/25/machine-2-machine-internet-of-things-real-world-internet/, 2011.
69. Joel Young, "Web Services Put M2M in the Cloud," http://www.eetimes.com/design/embedded/4219528/Web-services-puts-M2M-in-The-Cloud, 2011.
70. ZTE Corporation, "Opportunities, Challenges, and Practices of the Internet of Things," http://wwwen.zte.com.cn/endata/magazine/ztetechnologics/2010/no5/articles/201005/t20100510_184418.html, 2010.

71. B. Schilit, N. Adams, and R. Want, "Context-Aware Computing Applications," Proceedings of the 1994 First Workshop on Mobile Computing Systems and Applications, IEEE, 1994.
72. Richard MacManus, "DASH7: Bringing Sensor Networking to Smartphones," http://www.readwriteweb.com/archives/dash7_bringing_sensor_networking_to_smartphones.php, 2010.
73. Gartner Report, "Who's Who in Middleware," http://www-01.ibm.com/software/info/websphere/partners4/articles/gartner/garwho.html#fig1, 2004.
74. Honbo Zhou, *Smarter Earth: Deciphering Internet of Things* (Book in Chinese, http://baike.baidu.com/view/4114160.htm), Publishing House of Electronics Industry, 2010.
75. Honbo Zhou, *Cloud Computing: ICT's Tower of Babel* (Book in Chinese, http://baike.baidu.com/view/5276061.htm), Publishing House of Electronics Industry, 2011.
76. 3GPP Technical Reports, "Systems Improvements for Machine-Type Communications," http://www.3gpp.org/ftp/Specs/archive/23_series/23.888/, 2011.
77. Sahin Albayrak et al., "Smart Middleware for Mutual Service-Network Awareness in Evolving 3GPP Networks," http://pure.ltu.se/portal/files/2154720/04554377.pdf, 2008.
78. Jean-Marie Bonnin et al., "Mobile Wireless Middleware: Operating Systems and Applications," Proceedings of Mobilware, 2009.
79. Sasu Tarkoma, *Mobile Middleware: Architecture, Patterns and Practice*, Wiley, 2009.
80. Andreas Rasche, "Adaptive and Reflective Middleware," http://www.dcl.hpi.uni-potsdam.de/teaching/mds_07/mds10_adaptivemw.pdf.
81. Adam Dunkels et al., "IP for Smart Objects: Internet Protocol for Smart Objects (IPSO) Alliance,"http://www.sics.se/~adam/dunkels08ipso.pdf, 2008.
82. Mi Li et al., "Middleware for Sensor Network," http://www.eecg.toronto.edu/~jacobsen/courses/ece1770/slides/snetworks.ppt.
83. Miao-Miao Wang et al., "Middleware for Wireless Sensor Networks: A Survey," *Journal of Computer Science and Technology* 23(3): 305–326, May 2008.
84. Shuai Tong, "An Evaluation Framework for Middleware Approaches on Wireless Sensor Networks," http://www.cse.tkk.fi/en/publications/B/5/papers/tong_final.pdf.
85. Ralph Duncan, "A Survey of Parallel Computer Architectures," http://cs.nju.edu.cn/~gchen/teaching/fpc/Duncan90.pdf, 1990.

86. Honbo Zhou, "Distributed Computing of Weak and Strong Precedence Constrained Problems," Ph.D. Thesis, University of Zurich, Switzerland, 1993.
87. Honbo Zhou, "Parallel Architectures for Fast Image Processing," Proceedings of Conference on 3D Optical Measurement Techniques, Vienna, Austria, 1989.
88. Honbo Zhou and Lutz Richter, "Very Fast Distributed Spreadsheet Computing," *Journal of Systems and Software*, 25: 185–192, 1994.
89. Honbo Zhou, "Two-Stage M-Way Graph Partitioning," *Parallel Computing*, 19, 1359–1373, 1993.
90. Honbo Zhou, "Object Points Detection in a Photogrammetric Test Field," Proceedings of the ISPRS Commission V Symposium, Zurich, Switzerland, 1990.
91. Honbo Zhou, "Knowledge Based Parallel Recognition of Handwritten Alphanumerics," IEEE Proceedings of Intel. Conference on Acoustic, Speech, and Signal Processing, Glasgow, UK, 1989.
92. Honbo Zhou, "An Effective Approach for Distributed Program Allocation," *Journal of Parallel Algorithms and Applications,* 3: 57–71, 1993.
93. Honbo Zhou, "Scheduling DAGs on a Bounded Number Of Processors," Intel. Conference on Parallel and Distributed Processing: Techniques and Applications, Sunnyvale, 1996.
94. Honbo Zhou, "Image Processing in a Workstation-Based Distributed System," Proceedings of 2nd Intel. Conference on Automation, Robotics and Computer Vision, Singapore, 1992.
95. Honbo Zhou, "Enhancement and Delineation of Lung Tumors in Local X-ray Chest Images," SPIE Proceedings: Visual Communication and Image Processing, Lausanne, Switzerland, 1990.
96. "Top 500 Supercomputers," http://www.top500.org/. 2011.
97. R. Rocha, "Middleware for Location-based Services," http://www-di.inf.puc-rio.br/~endler/courses/Mobile/Monografias/04/Ricardo-Mono.pdf, 2004.
98. Henry Detmold et al., "Middleware for Video Surveillance Networks," http://dl.acm.org/citation.cfm?id=1176872, 2006.
99. Tiehan Lv et al., "Distributed Real-Time Embedded Video Processing," http://www.ll.mit.edu/HPEC/agendas/proc03/pdfs/lv.pdf.
100. Rogerio Feris et al., "Case Study: IBM Smart Surveillance System," http://rogerioferis.com/publications/FerisBookChapter09.pdf.

101. Vlad Trifa et al., "Web of Things: Connecting People and Objects on the Web," http://www.webofthings.com/sxsw/sxsw.pdf, 2010.
102. K. Jakobs et al., "Developing Standards for the IoT: A Collaborative Exercise!?" http://www.wi.rwth-aachen.de/Forschung/Developing%20Standards%20for%20the%20IoT.pdf, 2010.
103. IoT-A, "Internet of Things Architecture," http://www.iot-a.eu/public.
104. Inge Gronbaek, "M2M Architecture with Node and Topology Abstractions," Telektronikk, Feb. 2009.
105. Joachim Koss, "ETSI: M2M Activities in ETSI," http://ftp.tiaonline.org/GSC/GSC16/MSTF/20110920-21_AtlantaGA/Roundtable_Presentations/C%20GSC%20MSTF-Koss-ETSI.pdf, 2011.
106. W3C Incubator Group Report, "Semantic Sensor Network XG Final Report," http://www.w3.org/2005/Incubator/ssn/XGR-ssn-20110628/, 2011.
107. "Sensor Web Enablement," http://www.ogcnetwork.net/SWE.
108. SmartProducts, http://www.smartproducts-project.eu/, 2010.
109. SENSEI, http://www.ict-sensei.org/.
110. CASAGRAS Final Report, http://www.grifs-project.eu/data/File/Casagras_Final%20Report.pdf.
111. BRIDGE (Building Radio Frequency Identification for the Global Environment), http://www.bridge-project.eu/.
112. CUBIQ, http://www.cubiq.jp.
113. IoT-A, "Project Deliverable D1.2: Initial Architectural Reference Model for IoT," http://www.iot-a.eu/public/public-documents/d1.2. Ongoing.
114. E. Nordmark et al., "Shim6: Level 3 Multihoming Shim Protocol for IPv6," Request for Comments 5533, Internet Engineering Task Force, 2009.
115. "The Internet of Things," http://www.sensinode.com/.
116. "IEEE Standard for SCADA and Automation Systems," http://morse.colorado.edu/~tlen5830/ho/IEEE08C37_1.pdf, 2007.
117. "The Ultimate M2M Communication Protocol," http://www.bitxml.org/doc/BITXml_protocol_EN_2.0.1.pdf, 2007.
118. "OPC Unified Architecture: The Universal Communication Platform for Standardized Information Models," http://www.opcfoundation.org/DownloadFile.aspx/Brochures/OPC-UA-CollaborationOverview.pdf?RI=803.

119. Chris Anderson, "The Long Tail," http://www.wired.com/wired/archive/12.10/tail.html.
120. The OPC Foundation, http://www.opcfoundation.org/.
121. OSGi Alliance, http://www.osgi.org/.
122. M. Bosquet, "Gridwise Standards Mapping Overview," Pacific Northwest National Laboratory, Report PNNL-14587, 2004.
123. Kang Lee, "Sensor Standards Harmonization," http://ieee1451.nist.gov/Sensors_Harmonization/membersonly/SSH_WG_Meeting_March-14-2006/Sensor_Stds_Harmonization.pdf.
124. uID Center Web Site, "What Is Ucode?" http://www.uidcenter.org/learning-about-ucode/what-is-ucode.
125. Dialog Project, http://dialog.hut.fi/.
126. "IMEI Allocation and Approval Guidelines," http://www.algerietelecom.dz/veilletech/bulletin67/pdf/mobile5.pdf.
127. "Constrained Application Protocol (CoAP)," draft-ietf-core-coap-01, http://tools.ietf.org/html/draft-ietf-core-coap-01.
128. "Introduction to MHP and GEM," http://www.mhp.org/introduction.htm.
129. "M2M and SCADA Convergence," http://m2m.orangeom.com/m2m-and-scada-convergence/.
130. "Universal Middleware," http://soa.sys-con.com/node/492519?page=0,0.
131. The SODA Alliance, http://www.sensorplatform.org/soda/.
132. mHealth Summit, http://www.mhealthsummit.org/.
133. Hydra, http://www.hydramiddleware.eu/.
134. Chao Chen et al., "Device Integration in SODA Using the Device Description Language," http://www.icta.ufl.edu/dundee/DundeeFloridaExchange/ppt-chao.pdf.
135. "Network Enterprise Technology Command," http://www.globalsecurity.org/military/agency/army/asc.htm.
136. Tobias Heer et al., "Security Challenges in the IP-based Internet of Things," http://www.comsys.rwth-aachen.de/fileadmin/papers/2011/2011-heer-iot-challenges.pdf.
137. "Standardized M2M Software Development Platform," http://www.interdigital.com/images/id_misc/Standardized_M2M_SW_Dev_Platform.pdf.
138. Christopher Strachey, "Time Sharing in Large, Fast Computers," *IFIP Congress* 336–341,1959.

139. Honbo Zhou, Al Geist et al., "LPVM: A Step Towards Multithreaded PVM," *The Journal of Concurrency: Practice and Experience*, 10(5): 407–416, April 1998.
140. Honbo Zhou, "Faster (ATM) Message Passing in PVM," 9th Intel. Parallel Processing Symposium: Workshop on High-Speed Network Computing, Santa Barbara, 1995.
141. Honbo Zhou and A. Geist, "'Receiver Makes Right' Data Conversion in PVM," 14th Intel. Phoenix Conference on Computers and Communications, Phoenix, 1995.
142. Honbo Zhou, Joe Skovira et al., "The EASY: LoadLeveler API Project," 10th IPPS: Job Scheduling Strategies for Parallel Processing, Hawaii, 1996.
143. Jeffrey Dean and Sanjay Ghemawat, "MapReduce: Simplified Data Processing on Large Clusters," http://static.googleusercontent.com/external_content/untrusted_dlcp/labs.google.com/zh-CN//papers/mapreduce-osdi04.pdf, 2004.
144. Torsten Hoefler, Andrew Lumsdaine, and Jack Dongarra, "Towards Efficient MapReduce Using MPI," http://www.unixer.de/publications/img/hoefler-map-reduce-mpi.pdf.
145. Sanjay Ghemawat, Howard Gobioff, and Shun-Tak Leung, "The Google File System," 19th ACM Symposium on Operating Systems Principles, 2003.
146. Fay Chang, Jeffrey Dean, Sanjay Ghemawat et al., "Bigtable: A Distributed Storage System for Structured Data," Seventh Symposium on Operating System Design and Implementation, 2006.
147. Hadoop, http://hadoop.apache.org/core.
148. Dhruba Borthakur, "Hadoop and Condor," http://www.grid.org.il/_Uploads/dbsAttachedFiles/hadoop_condor.ppt.
149. "Condor Workers on Amazon EC2," http://www.isi.edu/~gideon/condor-ec2/, 2008.
150. IBM General Parallel File System (GPFS), http://www-03.ibm.com/systems/software/gpfs/.
151. Christian Engelmann, Hong Ong, and Stephen L. Scott, "Middleware in Modern High Performance Computing System Architectures," Lecture Notes in Computer Science: Proceedings of International Conference on Computational Science, Springer, 2007.
152. Thierry Priol, "Grid Middleware," http://gridatasia.ercim.eu/download/Beijing/22/Workshop-Middleware-Priol.pdf.

153. Domenico Talia et al., *Grid Middleware and Services: Challenges and Solutions*, Springer, 2008.
154. Peter Mell and Timothy Grance, "The NIST Definition of Cloud Computing," http://csrc.nist.gov/publications/nistpubs/800-145/SP800-145.pdf.
155. A. Anjomshoaa et al., "Job Submission Description Language (JSDL) Specification," Open Grid Forum, 2005.
156. A. Grimshaw, S. Newhouse et al., "OGSA Basic Execution Service Version 1.0," Open Grid Forum, 2006.
157. Jeremy Geelan, "The Top 250 Players in the Cloud Computing Ecosystem," http://cloudcomputing.sys-con.com/node/1386896.
158. Alex Salkever, "The Internet of Things and the Cloud," http://gigaom.com/cloud/alex-salkever-on-the-internet-of-things/.
159. ABI Research, "Mobile Cloud Computing," http://www.abiresearch.com/research/1003385-Mobile+Cloud+Computing, 2010.
160. Qusay H. Mahmoud, *Middleware for Communications*, John Wiley & Sons, 2003.
161. Paolo Bellavista and Antonio Corradi, editors, *The Handbook of Mobile Middleware*, John Wiley & Sons, 2006.
162. V. Reding, "Internet of the Future: Europe Must Be a Key Player," Speech to the Future of the Internet initiative of the Lisbon Council, Brussels, 2009.
163. Real World Internet (Internet of Things) Cluster of FIA, http://rwi.future-internet.eu/index.php/Main_Page.
164. "Parliament of Things," http://www.readwriteweb.com/archives/parliament_of_things.php.
165. Edward Lee, "Cyber Physical Systems: Design Challenges," University of California, Berkeley Technical Report No. UCB/EECS-2008-8.
166. "2010: The Year of the Strong Grid?" http://blogs.forbes.com/williampentland/2010/12/16/critical-infrastructure-dg/.
167. Eric Cheng, Internet of Things – China, http://www.finnode.fi/files/43/IoT_Eric_Cheng_09062011.pdf.
168. Chan Kim and Renee Mauborgne, "Blue Ocean Strategy," http://www.blueoceanstrategy.com/.
169. Thomas Friedman, "The World Is Flat," http://www.thomaslfriedman.com/bookshelf/the-world-is-flat.
170. ARTEMIS: Advanced Research and Technology for Embedded Intelligence and Systems, http://www.artemisia-association.org/.

171. R. Achatz of Siemens, quoted in *The Economist*, "Revving up: How globalisation and information technology are spurring faster innovation," http://www.economist.com/node/9928259, 2007.
172. http://energyperformancecontracting.org/.
173. http://www1.eere.energy.gov/femp/.
174. http://www.oe.energy.gov/smartgrid.htm.
175. "The Modern Grid Initiative," http://www.smartgridnews.com/artman/uploads/1/ModernGridInitiative_Final_v2_0.pdf, 2007.
176. Gary Locke et al., http://www.nist.gov/public_affairs/releases/upload/smartgrid_interoperability_final.pdf, 2010.
177. P. Gouvas, T. Bouras, and G. Mentzas, "An OSGi-based Semantic Service-Oriented Device Architecture," OTM Workshops, http://imu.ntua.gr/Papers/C92-OTM-PerSys-2007.pdf, 2007.
178. "The World's Smallest OSGi Solution," http://www.prosyst.com/index.php/de/html/news/details/18/smallest-OSGi/.
179. http://cs-people.bu.edu/gtw/motes/.
180. "The Planet Will Be Instrumented, Interconnected, Intelligent," http://www.ibm.com/smarterplanet/za/en/overview/visions/index.html.
181. "Digital TV Middleware, Standards and Trends," http://news.frbiz.com/digital_tv_middleware_standards-365778.html.
182. "The Future of Search and SEO (Location-based Search on Devices)," http://www.adambullas.com/news/2010/05/30/the-future-of-search-and-seo/.
183. "M2M Trend: Vertical Extension and Horizontal Convergence," http://wwwen.zte.com.cn/endata/magazine/ztetechnologies/2010/no7/articles/201007/t20100715_187385.html.
184. "M2M Service Platform to Support 'Carrier Cloud'," NEC Report, http://www.nec.co.jp/techrep/en/journal/g10/n02/100220.pdf.
185. "M2M: A New Age of Telemetry," http://www.metrilog.at/download/M2M_WhitePaper.pdf.
186. "Machine-To-Machine (M2M) & Smart Systems Forecast—2010-2014," Harbor Research's 2010 M2M & Smart Systems Forecast Report, 2010.
187. *M2M Magazine*, http://www.m2mmag.com/m2mnew/connected_world/agenda08.aspx.
188. "2010 M2M 100," http://www.m2mdatasmart.com/news/m2m-magazine-names-the-m2m-100-for-2010.html.

189. Alan Weissberger, "Exponential Growth in M2M Market Dependent on Important Network Enhancements," http://viodi.com/2010/10/07/exponential-growth-in-m2m-market-dependent-on-important-network-enhancements/.
190. "Top 10 Internet of Things Developments of 2010," http://www.readwriteweb.com/archives/top_10_internet_of_things_developments_of_2010.php.
191. "U-SNAP Alliance Launched to Extend Smart Grid to Energy Aware Consumer Products," http://www.automatedbuildings.com/releases/apr09/090429105828usnap.htm.
192. "Adidas miCoach, Nike+, Sensor Devices Get People Exercising," http://www.usatoday.com/news/health/weightloss/2010-01-28-workout28_st_N.htm.
193. M. Bosquet, "Gridwise Standards Mapping Overview," Pacific Northwest National Laboratory, Report PNNL-14587, 2004.
194. National Intelligence Council, "Six Technologies with Potential Impacts on US Interests out to 2025," http://www.fas.org/irp/nic/disruptive.pdf, 2008.
195. "Personal Navigation Devices: Making Money from a Declining Market," http://www.abiresearch.com/research/1006758-Personal+Navigation+Devices, 2011.
196. Ali Ipakchi, "Implementing the Smart Grid: Enterprise Information Integration," http://www.gridwiseac.org/pdfs/forum_papers/121_122_paper_final.pdf, 2007.
197. Frost & Sullivan Study, "Strategic Market and Technology Assessment of Telematics Applications for Electric Vehicles," http://www.cars21.com/files/papers/evs-telematics-frost-sullivan.pdf, 2010.
198. GTM Research, "Defining an End-to-End Smart Grid," http://www.greentechmedia.com/research/report/smart-grid-in-2010, 2009.
199. Pike Research, "Electric Vehicle Information Technology Systems," http://www.pikeresearch.com/research/electric-vehicle-information-technology-systems, 2010.
200. "IDC Smart Cities Index and Its Application in Spain," http://www.idc-ei.com/getdoc.jsp?containerId=EIRS56T, 2011.
201. "The Evolution of Cisco's Smart+Connected Communities," http://blogs.cisco.com/news/the_evolution_of_ciscos_smart connected_communities__to_colorado/, 2010.
202. "M2M Service Platform to Support 'Carrier Cloud,'" http://www.nec.co.jp/techrep/en/journal/g10/n02/100220.html, 2010.

203. "Operator Opportunities in the Internet of Things," http://www.ericsson.com/res/thecompany/docs/publications/ericsson_review/2011/er_edcp.pdf, 2011.
204. "Strategy Analytics: Enterprises Will Drive Mobile Device Management to a $5 Billion Market," http://goliath.ecnext.com/coms2/gi_0199-3454517/Strategy-Analytics-Enterprises-Will-Drive.html, 2004.
205. "M2M's Future Is Managing 2.1B Connected Things," http://www.billingworld.com/articles/2011/03/m2m-s-future-is-managing-2-1b-connected-things.aspx, 2010.
206. "Rise of the Machine-to-Machine: Wireless M2M Connection Market Booms," http://www.isuppli.com/Mobile-and-Wireless-Communications/News/Pages/Rise-of-the-Machine-to-Machine-Wireless-M2M-Connection-Market-Booms.aspx, 2010.
207. Press Release, "Number of Embedded Mobile & M2M Connected Devices to Rise to 412 Million Globally by 2014, says Juniper Research," http://juniperresearch.com/viewpressrelease.php?pr=178, 2010.
208. "M2M Value Chain," http://www.mincom.tn/fileadmin/PDF/Presentations/M2M_Tunisiana_WS.pdf, 2011.
209. "ABI Research Publishing Cellular M2M Module Vendor Market Shares for 2010," http://www.abiresearch.com/research_blog/1605, 2011.
210. "China's RFID Market Set to Double by 2014," http://www.isuppli.com/china-electronics-supply-chain/news/pages/chinas-rfid-market-set-to-double-by-2014.aspx, 2011.
211. "Wireless Sensor Networks 2011–2021," http://www.idtechex.com/research/reports/wireless-sensor-networks-2011-2021-000275.asp?viewopt=desc, 2011.
212. Barbara Pareglio, "Overview of ETSI M2M Architecture," http://docbox.etsi.org/Workshop/2011/201110_M2MWORKSHOP/02_M2M_STANDARD/M2MWG2_Architecture_PAREGLIO.pdf, 2011.
213. "Numerex DNA®," http://www.numerex.com/inside-our-DNA.
214. "Expert Voice: Paul Saffo on Smart Sensors," http://www.cioinsight.com/c/a/Expert-Voices/Expert-Voice-Paul-Saffo-on-Smart-Sensors/, 2002.
215. Chetan Sharma, "Managing Growth and Profits in the Yottabyte Era," http://mobilebroadbandopportunities.com/chetansharma/Sharma1.pdf, 2010.

216. Carlos Ralli Ucendo, "IPv6 Services in LONG Network," http://long.ccaba.upc.es/long/050Dissemination_Activities/IPv6_Services_LONG.pdf.
217. Igor Bélai et al., "The Industrial Communication Systems Profibus and Profinet," http://www.profibus.sk/uploads/media/Belai-Drahos_-_The_Industrial_Communication_Systems_PROFIBUS_and_PROFInet_01.pdf, 2009.
218. "Industrial Networks for Communication and Control," http://anp.tu-sofia.bg/djiev/PDF%20files/Industrial%20Networks.pdf.
219. "Common Industrial Protocol," http://www.technologyuk.net/telecommunications/industrial_networks/cip.shtml.
220. "Evolving Wireless Standards," http://cp.literature.agilent.com/litweb/pdf/5989-5539EN.pdf.
221. "Broadband Radio Access Networks," http://easy.intranet.gr/H2_RA.pdf, ETSI Technical Report, 1999.
222. "Industrial Automation Technologies," http://www.globalspec.com/reference/13923/121073/chapter-11-3-2-industrial-automation-technologies-the-erp-layer.
223. Bill McBeath, "Who Will Provide the 'Location' In Location-Based Services?" http://www.clresearch.com/research/detail.cfm?guid=CADA9F0D-3048-79ED-9930-134BF9519AAE, 2010.
224. "M2M Value Chain," http://www.wireless-technologies.eu/index.php?page=m2m-value-chain.
225. Long Nguyen Hoang, "Middlewares for Home Monitoring and Control," http://www.tml.tkk.fi/Publications/C/23/papers/NguyenHoang_final.pdf, 2007.
226. "Seriously Smart Software for M2M Data," http://m2m.tmcnet.com/topics/m2mevolution/articles/198791-seriously-smart-software-m2m-data.htm, 2011.
227. M2M Evolution Conference, http://m2m.tmcnet.com//conference/, Austin, Texas, October 3, 2012.
228. "RFID Middleware Is Extinct: The Intelligent Sensor Network Is Born," http://rfid.net/basics/middleware/143-rfid-middleware-is-extinct-the-intelligent-sensor-network-is-born, 2011.
229. "Agilla: Mobile Agent Middleware for Wireless Sensor Networks," http://www.cse.wustl.edu/~lu/cse521s/Slides/agilla.pdf.
230. ETSI, "Machine to Machine Communications," http://www.etsi.org/WebSite/document/EVENTS/ETSI%20M2M%20Presentation%20during%20MWC%202011.pdf, 2011.

231. "Middleware Technology for Digital Home Services," http://hometoys.com/emagazine.php?url=/ezine/08.04/perumal/middleware.htm.
232. "Magic Quadrant for Enterprise Application Servers," http://www.gartner.com/technology/reprints.do?id=1-17GUO5Z&ct=110928&st=sb=2011.
233. "Semantic Sensor Network XG Final Report," http://www.w3.org/2005/Incubator/ssn/XGR-ssn-20110628/, 2011.
234. "OPC Web Services," http://advosol.com/c-2-opc-xml-webservices.aspx.
235. OMA (Open Mobile Alliance), "OMA M2M Activities," http://docbox.etsi.org/workshop/2010/201010_M2MWORKSHOP/06_M2MGlobalCollaboration/LEUCA_OMABOARDx.pdf.
236. "Overall FI-WARE Vision," http://forge.fi-ware.eu/plugins/mediawiki/wiki/fiware/index.php/Overall_FI-WARE_Vision.
237. "Security Technologies for NGN," http://wwwen.zte.com.cn/endata/magazine/ztecommunications/2007year/no4/articles/200712/t20071224_162457.html.
238. "Secure Middleware for Embedded Peer to Peer Systems," http://www.ist-world.org/ProjectDetails.aspx?ProjectId=e599578c3b9949ae841501eb790e91b0&SourceDatabaseId=7cff9226e582440894200b751bab883f, 2009.
239. "Grand Challenge Application Experiences with PVM," http://www.netlib.org/utk/papers/comp-phy7/node11.html, 1996.
240. "An Introduction to Hyper-V in Windows Server 2008," http://technet.microsoft.com/en-us/magazine/2008.10.hyperv.aspx, 2008.
241. "Cloud Computing and SOA Innovations," https://www.ibm.com/developerworks/mydeveloperworks/blogs/zhanglj/entry/trend_4_open_standards_moving?lang=en,2007.
242. "VAMOS: Virtualization Aware Middleware," http://www.mulix.org/pubs/misc/vamos.pdf.
243. "Grid Resource Management: Challenges, Approaches, & Solutions," http://ww2.cs.mu.oz.au/678/GridRM.ppt.
244. "Cloud Middleware Market Shares, Strategies, and Forecasts, Worldwide, 2011 to 2017," http://www.researchmoz.com/cloud-middleware-market-shares-strategies-and-forecasts-worldwide-2011-to-2017-report.html, 2011.

245. "Multi-Tenant Data Architecture," http://msdn.microsoft.com/en-us/library/aa479086.aspx, 2006.
246. "Develop and Deploy Multi-tenant Web-delivered Solutions Using IBM Middleware," http://www.ibm.com/developerworks/webservices/library/ws-multitenantpart2/, 2009.
247. "Magic Quadrant for Application Infrastructure for Systematic SOA-Style Application Projects," http://www.juvo.be/en/blog/oracle-leader-all-3-gartner-magic-quadrants-soa-and-soa-governance, 2010.
248. Mitsutaka Itoh et al., "Virtual Smartphone over IP," https://www.ntt-review.jp/archive/ntttechnical.php?contents=ntr201007sf4.pdf&mode=show_pdf.
249. "Android Cloud to Device Messaging Framework," http://code.google.com/intl/en/android/c2dm/index.html.
250. "Fiberlink Launches Mobility as a Service," http://www.businesswire.com/news/home/20081016005554/en/Fiberlink-Launches-Mobility-Service-Platform, 2008.
251. "Pachube Extreme Connectivity," http://assets.en.oreilly.com/1/event/51/Extreme%20Connectivity%20Presentation.pdf.
252. "Internet Trends," http://www.morganstanley.com/institutional/techresearch/pdfs/MS_Internet_Trends_060710.pdf, 2010.
253. "Mobile Cloud Computing Opportunities and Challenges," http://cloudcomputingtopics.com/2010/11/mobile-cloud-computing-opportunities-and-challenges-for-mnos/, 2011.
254. C. Zhang et al., "Building a Smart Community Using [ez]IBS," *Journal of Dalian University*, 28(6), December 2007.
255. Honbo Zhou, "Unified Middleware for IoT applications," Keynote Address at The 1st International Workshop on Internet of Things Applications, http://eceweb1.rutgers.edu/~yyzhang/iot12/iot-pgm-final.pdf, 2012.
256. Thomas Friedman, "So Much Fun. So Irrelevant," http://www.nytimes.com/2012/01/04/opinion/friedman-so-much-fun-so-irrelevant.html?_r=2&ref=thomaslfriedman, 2012.
257. Chris Anderson, "Why the Internet of Things Finally Makes Sense," http://wsnblog.com/2011/11/15/why-the-internet-of-things-finally-makes-sense/, 2011.
258. Eugene Marinelli, "Hyrax: Cloud Computing on Mobile Devices Using MapReduce," http://reports-archive.adm.cs.cmu.edu/anon/2009/CMU-CS-09-164.pdf, 2009.

259. "Machina Research's Ten Predictions for M2M in 2012," http://www.m2mforumeurope.com/uploadedFiles/EventRedesign/UK/2012/June/19955002/Assets/Machina-Research-Ten-Predictions-for-M2M-in-2012-extract-for-M2M-Forum-Europe.pdf, 2012.
260. "M2M White Paper: The Growth of Device Connectivity," The FocalPoint Group, www.thefpgroup.com, 2010.
261. "The Internet of Things," IBM Video, http://www.youtube.com/watch?feature=player_embedded&v=sfEbMV295Kk, 2010.
262. "How the Internet of Things Will Change Everything—Including Ourselves," CISCO Video, http://www.youtube.com/watch?v=mf7HxU0ZR_Q&feature=related, 2011.
263. "Fleet and Asset Management Report 2012," Telematics Update, http://analysis.telematicsupdate.com/.
264. "Protocol Blenders and Information Creators: Middleware Providers and Implementation Strategy," http://www.automatedbuildings.com/news/dec08/articles/sinopoli/081129021941sinopoli.htm, 2008.
265. "Machine 2 Machine—Internet of Things—Real World Internet," http://machine2twomachine.wordpress.com/2011/08/25/machine-2-machine-internet-of-things-real-world-internet/.
266. S. Bandyopadhyay et al., "Role of Middleware for Internet of Things," http://www.scribd.com/doc/64235401/Role-Of-Middleware-For-Internet-Of-Things-A-Study, 2011.
267. "Adopting Cloud Computing: Enterprise Private Clouds," http://www.infosys.com/infosys-labs/publications/infosyslabs-briefings/documents/cloud-computing-enterprise-private-clouds.pdf, 2009.
268. "Planning Guide for Infrastructure as a Service (IaaS)," http://blogs.technet.com/b/privatecloud/archive/2012/04/05/planning-guide-for-infrastructure-as-a-service-iaas.aspx, 2012.
269. Forrester Research, Global Extended Internet Forecast, 2006–2012, September 2006.
270. ABI Research, RFID Market Update, 2006.
271. C.G. Bell, R. Chen and S. Rege, "The Effect of Technology on Near Term Computer Structures," *Computer*, 2 (5) 29–38, March/April 1972.
272. Mark Weiser, "The Computer for the Twenty-First Century," *Scientific American*, 1991.

273. Mario Gerla and Leonard Kleinrock, "Vehicular networks and the future of the mobile Internet," http://nrlweb.cs.ucla.edu/publication/show/702.
274. D. Washburn and U. Sindhu, "Helping CIOs Understand 'Smart City' Initiatives", *Forrester*, 2010.
275. Andrew Brown, "A Brave New World in Mobile Machine to Machine (M2M) Communications," July 2008.
276. R. Chellappa, "Intermediaries in Cloud-Computing: A New Computing Paradigm", INFORMS Annual Meeting, Dallas, TX, October 1997.
277. Eric Schmidt, "Conversation with Eric Schmidt hosted by Danny Sullivan" Search Engine Strategies Conference, August 9, 2006, http://www.google.com/press/podium/ses2006.html.
278. Ian Foster et al., "Cloud Computing and Grid Computing 360-Degree Compared," IEEE Grid Computing Environments Workshop, 2008.
279. Nicholas Carr, *The Big Switch: Rewiring the World, from Edison to Google*, New York: Norton, 2008.

Index

Note: page numbers in bold represent figures or tables.

C

F

H

J

K

L

M

O

P

Q

S

T

W

Y

Z

Lightning Source UK Ltd.
Milton Keynes UK
UKHW02n0642190818
327436UK00008B/95/P